Everyday Mathematics®

The University of Chicago School Mathematics Project

Student Math Journal
Volume 2

The McGraw·Hill Companies

The University of Chicago School Mathematics Project (UCSMP)

Max Bell, Director, UCSMP Elementary Materials Component; Director, *Everyday Mathematics* First Edition
James McBride, Director, *Everyday Mathematics* Second Edition
Andy Isaacs, Director, *Everyday Mathematics* Third Edition
Amy Dillard, Associate Director, *Everyday Mathematics* Third Edition

Authors

Max Bell
Jean Bell
John Bretzlauf
Amy Dillard
Robert Hartfield
Andy Isaacs
James McBride
Cheryl G. Moran*
Kathleen Pitvorec
Peter Saecker

**Third Edition only*

Technical Art

Diana Barrie

Teachers in Residence

Kathleen Clark, Patti Satz

Editorial Assistant

John Wray

Contributors

Robert Balfanz, Judith Busse, Mary Ellen Dairyko, Lynn Evans, James Flanders, Dorothy Freedman, Nancy Guile Goodsell, Pam Guastafeste, Nancy Hanvey, Murray Hozinsky, Deborah Arron Leslie, Sue Lindsley, Mariana Mardrus, Carol Montag, Elizabeth Moore, Kate Morrison, William D. Pattison, Joan Pederson, Brenda Penix, June Ploen, Herb Price, Dannette Riehle, Ellen Ryan, Marie Schilling, Susan Sherrill, Patricia Smith, Robert Strang, Jaronda Strong, Kevin Sweeney, Sally Vongsathorn, Esther Weiss, Francine Williams, Michael Wilson, Izaak Wirzup

Photo Credits

©Fotosearch, p. v *bottom*; Getty Images, cover, *right;* ©Linda Lewis; Frank Lane Picture Agency/CORBIS, cover, *bottom left;* ©Photodisc/Getty Images, p. v *top;* ©Star/zefa/Corbis, cover, *center*.

---

www.WrightGroup.com

Printed in the United States of America.

Send all inquiries to:
Wright Group/McGraw-Hill
P.O. Box 812960
Chicago, IL 60681

ISBN 0-07-604555-2

16 CPC 12 11 10

Contents

UNIT 7 Patterns and Rules

Using a Calculator to Find Patterns 161
Math Boxes 7◆1 162
Making 10s 163
Solving Subtraction Problems 164
Math Boxes 7◆2 165
Playing *Basketball Addition* 166
Basketball Addition 167
More Multiplication Number Stories 168
Math Boxes 7◆3 169
The Wubbles 170
Math Boxes 7◆4 171
Math Boxes 7◆5 172
Record of Our Jumps 173
Record of Our Arm Spans 174
Math Boxes 7◆6 175
The Lengths of Objects 176
Math Boxes 7◆7 178
Soccer Spin Directions 179
Table of Our Arm Spans 180
Bar Graph of Our Arm Spans 181
Math Boxes 7◆8 182
Math Boxes 7◆9 183

UNIT 8 Fractions

Equal Parts 184
Dressing for School 185
Math Boxes 8◆1 186
Pattern-Block Fractions 187
Geoboard Fences 189
Math Boxes 8◆2 190
Equal Shares 191
Fractions of Sets 192
Equal Parts 193
Math Boxes 8◆3 194
Math Boxes 8◆4 195
Equivalent Fractions 196
Equivalent Fractions Game Directions 198
Fractions of Collections 200
Math Boxes 8◆5 202
Fraction Top-It Directions 203
Math Boxes 8◆6 205
Fraction Number Stories 206
Math Boxes 8◆7 207
Math Boxes 8◆8 208

UNIT 9 Measurement

Yards 209
Possible Outcomes 210
Math Boxes 9•1 211
Units of Linear Measure 212
Math Boxes 9•2 213
Measuring Lengths with a Ruler 214
Math Boxes 9•3 215
Distance Around and Perimeter 216
Math Boxes 9•4 217
Driving in the West 218
Addition and Subtraction Practice 219
Math Boxes 9•5 220
Math Message 221
Math Boxes 9•6 222
Math Boxes 9•7 223
Equivalent Units of Capacity 224
Math Boxes 9•8 225
Weight 226
Math Boxes 9•9 227
Math Boxes 9•10 228

UNIT 10 Decimals and Place Value

Math Boxes 10•1 229
Good Buys Poster 230
Ways to Pay 231
Word Values 232
Math Boxes 10•2 233
Calculator Dollars and Cents 234
Pick-a-Coin Directions 236
Pick-a-Coin Record Tables 237
Finding the Median 238
Math Boxes 10•3 239
Then-and-Now Poster 240
Then-and-Now Prices 241
Math Boxes 10•4 242
Estimating and Buying Food 243
Math Boxes 10•5 244
Math Boxes 10•6 245
Making Change 246
The Area of My Handprint 248
The Area of My Footprint 249
Worktables 250
Geoboard Dot Paper 251
Math Boxes 10•7 252
Money Exchange Game Directions 253
Place-Value Mat 254
Ballpark Estimates 255
Math Boxes 10•8 256
Math Boxes 10•9 257
Place Value 258
Math Boxes 10•10 259
Parentheses Puzzles 260
Math Boxes 10•11 261
Math Boxes 10•12 262

UNIT 11 Whole Number Operations Revisited

Math Boxes 11•1 263
Art Supply Poster 264
Buying Art Supplies 265
Comparing Costs 266
Data Analysis 267
Math Boxes 11•2 268
Trade-First Subtraction 269
Math Boxes 11•3 270
Math Boxes 11•4 271
Multiplication Number Stories 272
Division Number Stories 274
Math Boxes 11•5 276
Multiplication Facts List 277
Using Arrays to Find Products 278
Math Boxes 11•6 279
Products Table 280
Math Boxes 11•7 281
Multiplication/Division Fact Families 282
Multiplication and Division with 2, 5, and 10 . . . 283
Math Boxes 11•8 284
Math Boxes 11•9 285
Beat the Calculator 286
Math Boxes 11•10 288

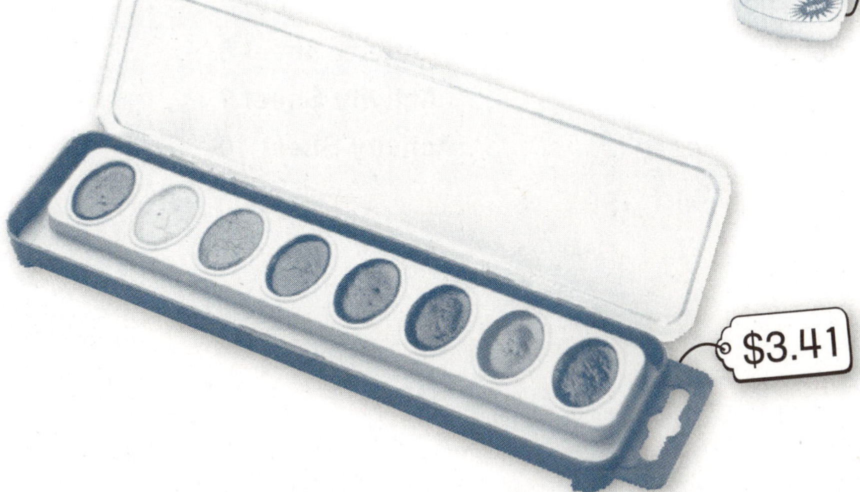

UNIT 12 Year-End Reviews and Extensions

Review: Telling Time 289
Math Boxes 12•1 290
Time Before and After 291
Many Names for Times 292
Addition and Subtraction Strategies 293
Math Boxes 12•2 294
Important Events in Communication 295
Math Boxes 12•3 296
Subtraction Practice 297
Math Boxes 12•4 298
Related Multiplication and Division Facts 299
Addition Card Draw Directions 300
Math Boxes 12•5 301
Animal Bar Graph 302
Interpreting an Animal Bar Graph 303
Math Boxes 12•6 304
Height Changes 305
Math Boxes 12•7 308
Math Boxes 12•8 309
Table of Equivalencies 310

Activity Sheets

Fraction Cards . . . . **Activity Sheet 5**
Fraction Cards . . . . **Activity Sheet 6**
×, ÷ Fact Triangles 1 . . . . **Activity Sheet 7**
×, ÷ Fact Triangles 2 . . . . **Activity Sheet 8**
×, ÷ Fact Triangles 3 . . . . **Activity Sheet 9**
×, ÷ Fact Triangles 4 . . . . **Activity Sheet 10**

Date Time

Using a Calculator to Find Patterns

1. Use a calculator to count by 5s starting with the number 102. Color the counts on the grid with a crayon. Look for a pattern.

									100
101	102	103	104	105	106	107	108	109	110
111	112	113	114	115	116	117	118	119	120
121	122	123	124	125	126	127	128	129	130

2. Pick a number to count by. Start with a number less than 310. Use your calculator to count. Record your counts on the grid with a crayon.

									300
301	302	303	304	305	306	307	308	309	310
311	312	313	314	315	316	317	318	319	320
321	322	323	324	325	326	327	328	329	330
331	332	333	334	335	336	337	338	339	340
341	342	343	344	345	346	347	348	349	350
351	352	353	354	355	356	357	358	359	360
361	362	363	364	365	366	367	368	369	370

I counted by ______ starting with the number ______.

Here is a pattern that I found: ________________________________

____

____

Date Time

LESSON 7•1

Math Boxes

1. Which one is certain to happen? Circle the best answer.

   A. A spaceship will land at school.

   B. Your favorite sports team will win every time.

   C. Spring will follow winter.

   D. You will be a movie star.

2. Solve.

   Unit

   17 − 9 = ______

   27 − 9 = ______

   57 − 9 = ______

   ______ = 77 − 9

   ______ = 97 − 9

3. Make a 7-by-7 array with dots.

   How many in all? ______ dots

4. Arrange the allowances in order from the minimum (smallest) to the maximum (largest).

   \$10, \$3, \$7, \$1, \$4

   _____, _____, _____, _____, _____

   The minimum is ______.

   The maximum is ______.

5. Match each person with the correct weight.

newborn	about 144 pounds
2nd grader	about 63 pounds
adult	about 7 pounds

6. How many boxes are on this Math Boxes page?

   ______ boxes

   How many boxes are on $\frac{1}{2}$ of this page?

   ______ boxes

Date Time

Making 10s

Record three rounds of *Hit the Target.*

Example Round:

Target number: 40

Starting Number	Change ⤻	Result	Change ⤻	Result	Change ⤻	Result
12	+38	50	−10	40		

Round 1

Target number: ______

Starting Number	Change ⤻	Result	Change ⤻	Result	Change ⤻	Result

Round 2

Target number: ______

Starting Number	Change ⤻	Result	Change ⤻	Result	Change ⤻	Result

Round 3

Target number: ______

Starting Number	Change ⤻	Result	Change ⤻	Result	Change ⤻	Result

Date Time

Solving Subtraction Problems

Use base-10 blocks to help you subtract.

1.

longs	cubes
5	6
− 3	9

2.

longs	cubes
7	3
− 1	4

Use any strategy to solve.

3. Ballpark estimate:

____

$$\begin{array}{r} 47 \\ -19 \\ \hline \end{array}$$

4. Ballpark estimate:

____

$$\begin{array}{r} 88 \\ -23 \\ \hline \end{array}$$

5. Ballpark estimate:

____

$$\begin{array}{r} 82 \\ -65 \\ \hline \end{array}$$

6. Ballpark estimate:

____

$$\begin{array}{r} 64 \\ -38 \\ \hline \end{array}$$

Math Boxes

1. 24 children. 6 in each row. Draw an array.

How many rows? _____ rows

How many children left over?

_____ children

MRB 112 113

2. 15 dogs.
13 cats.
12 birds.

How many animals?

_____ animals

3. Fill in the missing numbers.

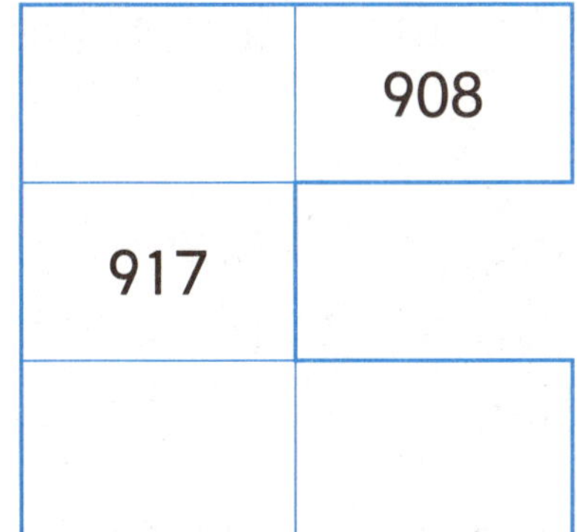

4. Draw a line of symmetry on this triangle.

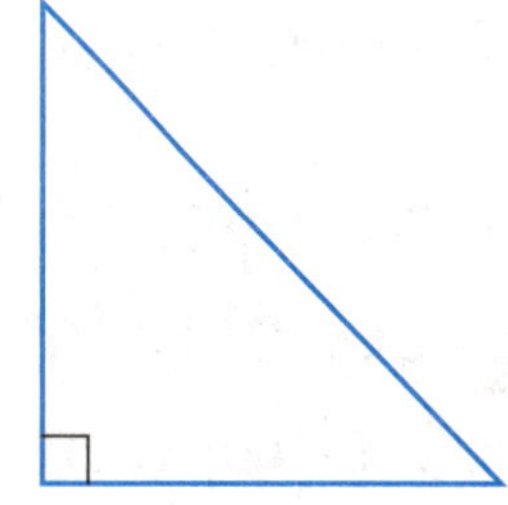

5. Find the rule. Complete the table.

in	out
193	183
232	222
441	
	346

6. _____ of the shape is shaded. Circle the best answer.

A. 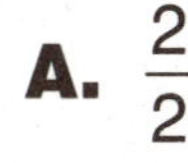

B. 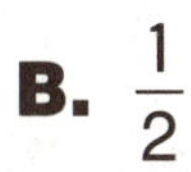

C. 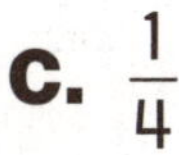

D. 0

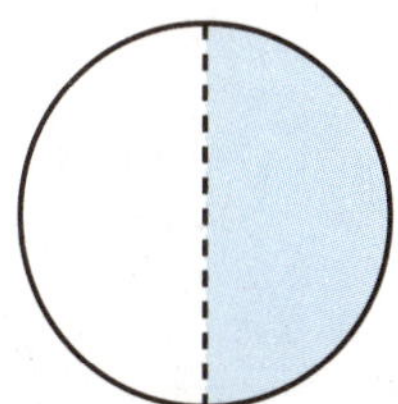

MRB 13

LESSON 7•3

Playing *Basketball Addition*

Materials
- ☐ *Basketball Addition* scoreboard (*Math Journal 2,* p. 167 or *Math Masters,* p. 451)
- ☐ 3 regular dice

Players 2 teams of 3–5 players each

Skill Add three or more 1- and 2-digit numbers

Object of the Game To score a greater number of points

Directions

1. Players on opposite teams take turns rolling the 3 dice.
2. Each player enters the sum of the numbers on the 3 dice in the Points Scored table.
3. After each player on a team has rolled the dice, each team finds the total number of points scored by their team for the first half of the game and enters the Team Score in the table.
4. Players repeat Steps 1–3 to find their team's score for the second half of the game.
5. Each team adds their team totals from both halves of the game to find their team's final score.
6. The team with the greater number of points wins the game.

Date Time

Basketball Addition

	Points Scored			
	Team 1		Team 2	
	1st Half	2nd Half	1st Half	2nd Half
Player 1				
Player 2				
Player 3				
Player 4				
Player 5				
Team Score				

Point Totals	**1st Half**	**2nd Half**	**Final**
Team 1	______	______	______
Team 2	______	______	______

1. Which team won the first half? ________________

   By how much? ______ points

2. Which team won the second half? ________________

   By how much? ______ points

3. Which team won the game? ________________

   By how much? ______ points

More Multiplication Number Stories

Write your own multiplication stories and draw pictures of your stories. You can use the pictures at the side of the page for ideas.

For each story:

- Write the words.
- Draw a picture.
- Write the answer.

There are 5 tricycles. How many wheels in all?

Answer: *15 wheels*
(unit)

1. ______________________________

____

Answer: ______________
(unit)

2. ______________________________

____

Answer: ______________
(unit)

A person has 2 ears.

A tricycle has 3 wheels.

A car has 4 wheels.

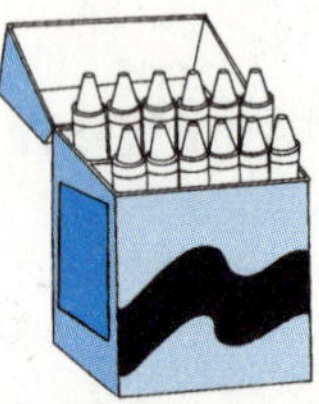

The box has 12 crayons.

The box has 100 paper clips.

The juice pack has 6 cans.

Date Time

LESSON 7•3

Math Boxes

1. Which one is impossible? Choose the best answer.

   ⬭ I will write a letter to a friend.

   ⬭ I will eat a piece of cake.

   ⬭ The sun will shine.

   ⬭ A fish will live out of water.

2. Solve.

   Unit

   ______ = 37 + 9

   ______ = 137 + 9

   116 − 8 = ______

   176 − 8 = ______

3. Draw 5 fish bowls and 2 fish in each bowl.

   How many fish in all?

   ______ fish

4. Arrange the number of pets in order from the minimum (smallest) to the maximum (largest).

   7, 0, 4, 1, 3, 5, 2

   ___, ___, ___, ___, ___, ___, ___

   The maximum is ______.

   The minimum is ______.

5. 1 bag of sugar weighs 5 pounds.

   6 bags of sugar weigh

   ______ pounds.

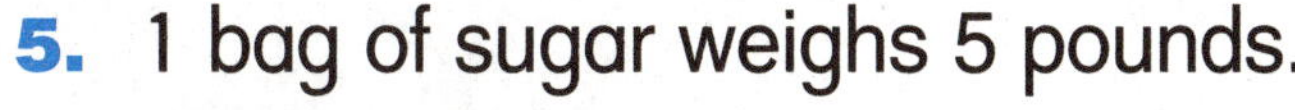

   A. 50 **B.** 60

   C. 25 **D.** 30

6. Circle the trapezoid that has $\frac{1}{3}$ shaded.

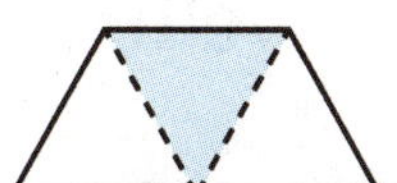

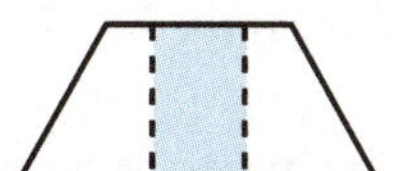

LESSON 7•4

The Wubbles

1. On each line, write the number of Wubbles after doubling. Use your calculator to help you.

You started on Friday with ______ Wubble.

On Saturday, there were ______ Wubbles.

On Sunday, there were ______ Wubbles.

On Monday, there will be ______ Wubbles.

On Tuesday, there will be ______ Wubbles.

On Wednesday, there will be ______ Wubbles.

On Thursday, there will be ______ Wubbles.

On Friday, there will be ______ Wubbles.

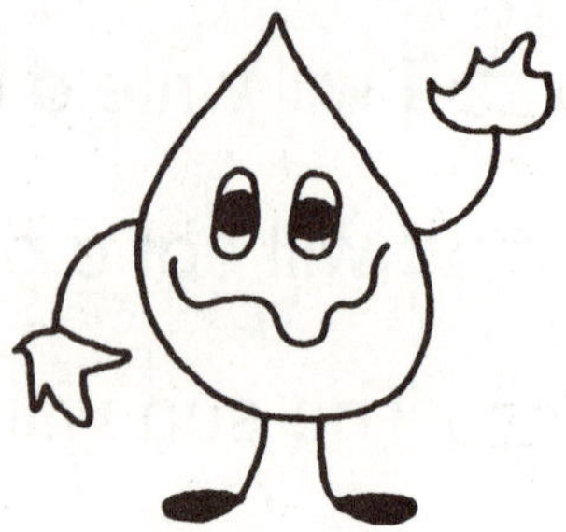

A Wubble

2. On each line, write the number of Wubbles after halving. Use your calculator to help you. Remember that "$\frac{1}{2}$ of" means "divide by 2."

There were ______ Wubbles.

After Wink 1, there were ______ Wubbles.

After Wink 2, there were ______ Wubbles.

After Wink 3, there were ______ Wubbles.

After Wink 4, there were ______ Wubbles.

After Wink 5, there were ______ Wubbles.

After Wink 6, there were ______ Wubbles.

After Wink 7, there was ______ Wubble.

Your room could look like this! What will you do?

Adapted with permission from *Calculator Mathematics Book 2* by Sheila Sconiers, pp. 10 and 11 (Everyday Learning Corporation, © 1990 by the University of Chicago).

Date Time

LESSON 7•4

Math Boxes

1. Collect 29 counters. How many groups of 3 can you make?

____ groups

How many counters are left over?

____ counters

2. Solve.

Unit
train cars

4 + 3 + 13 = ____

____ = 12 + 6 + 8

5 + 4 + 18 = ____

____ = 18 + 12 + 6

40 = 15 + 6 + ____

3. Fill in the missing numbers.

	717	

4. Draw the lines of symmetry on this rectangle.

How many lines of symmetry are there? ____

5. **Rule**
Double

in	out
2	4
4	
5	
	14

6. Shade one half of this square.

LESSON 7•5

Math Boxes

1. Draw hands to show 7:15.

2.

Recess Time

Number of Minutes

30
25
20
15
10
5
0

Mon Tue Wed Thu Fri

Which day had the longest recess? Circle the best answer.

A. Tuesday **B.** Friday

C. Wednesday **D.** Monday

3. Use counters to make a 5-by-3 array. Draw the array.

How many counters in all?

_____ counters

4. Fill in the missing numbers.

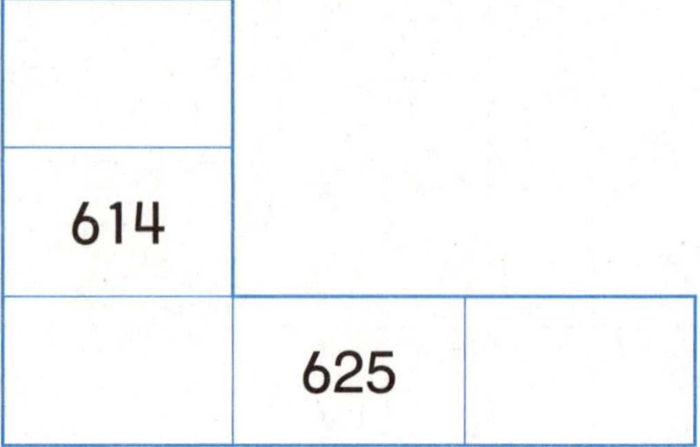

5. Draw or write the names of two things in the classroom that are the shape of a rectangular prism.

6. Write the fraction.

The part shaded = _____

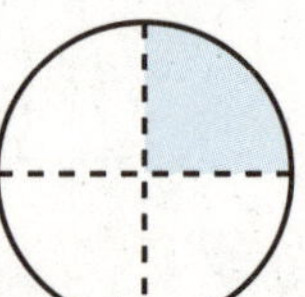

LESSON 7•6

Record of Our Jumps

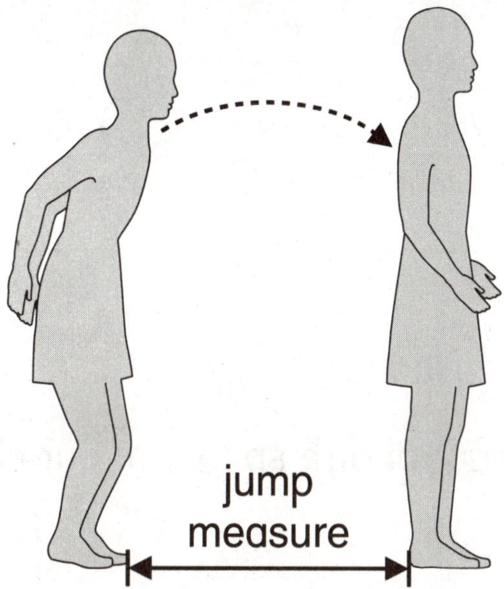

Place a penny or other marker (or make a dot with chalk) where the Jumper's back heel lands. Measure from the starting line to the marker. The jumps are measured to the nearest centimeter.

1. Record two of your jumps. Measure jumps to the nearest centimeter.

   First try: ______ centimeters

   Second try: ______ centimeters

2. My longer jump was ______ centimeters.

3. A middle value of jumps for our class is ______ centimeters.

Date Time

Record of Our Arm Spans

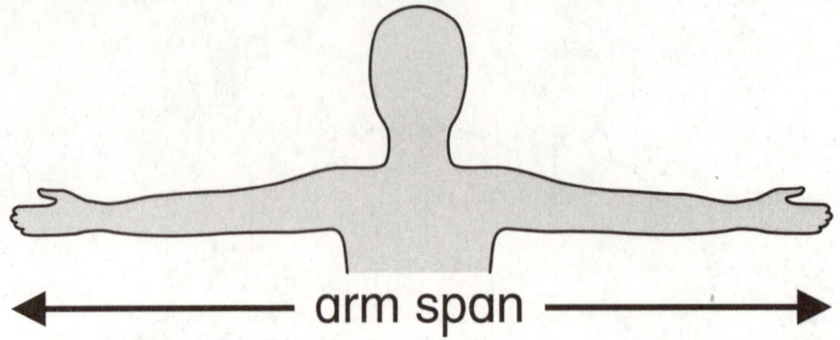

1. My arm span is ______ inches.

2. A middle value (median) of arm spans for our class is ______ inches.

Date Time

LESSON 7•6

Math Boxes

1. Show five possible ways to make 40¢.

2. 2 bunches of bananas. Each bunch has 5 bananas. How many bananas in all?

_______ bananas

Complete the diagram.

bunches	bananas per bunch	bananas in all

3. What is the temperature? Fill in the circle next to the best answer.

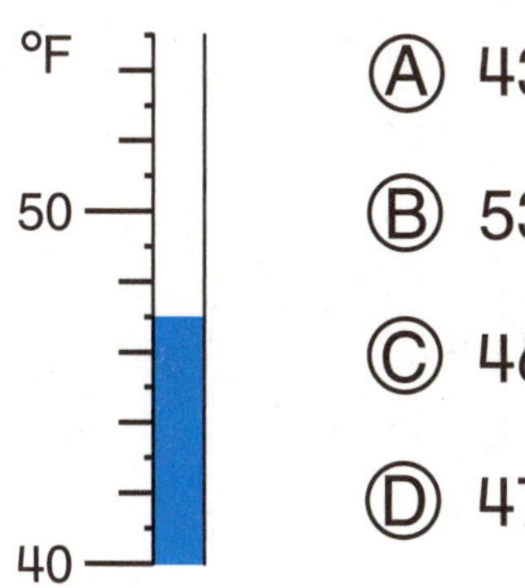

Ⓐ 43°
Ⓑ 53°
Ⓒ 46°
Ⓓ 47°

4. Which number is the most popular (the mode)?

Unit
minutes of homework

____

12, 12, 12, 14, 15, 16, 16

5. Make a ballpark estimate. Then solve the problem.
Ballpark estimate:

____

$$\begin{array}{r} 52 \\ -\ 29 \\ \hline \end{array}$$

6. Write <, >, or = in the box.

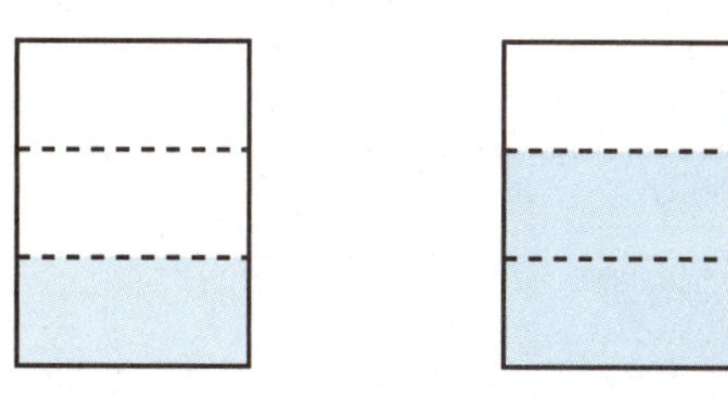

$\frac{1}{3}$ □ $\frac{2}{3}$

Date Time

LESSON 7•7

The Lengths of Objects

Reminder: *in.* means *inches; cm* means *centimeters*

Measure each item to the nearest inch.
Measure each item to the nearest centimeter.
Record your answers in the blank spaces.

1. **pencil**

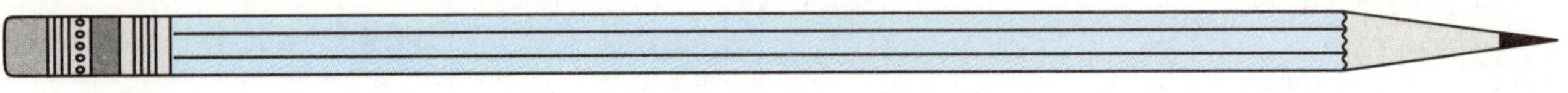

about ______ in.

about ______ cm

2. **screwdriver**

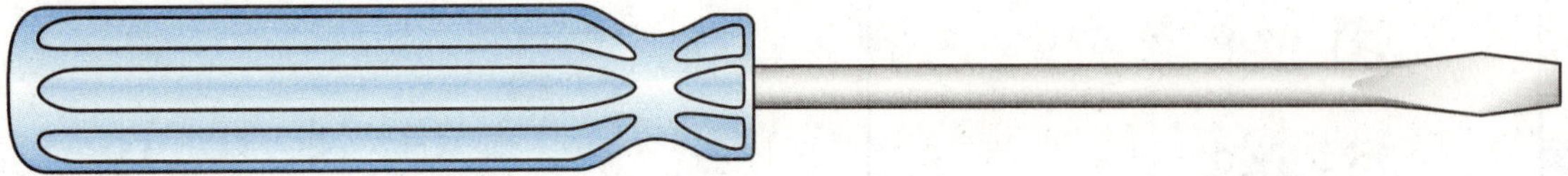

about ______ in.

about ______ cm

3. **pen**

about ______ in.

about ______ cm

The Lengths of Objects *continued*

4. bolt

about ______ in.

about ______ cm

5. dandelion leaf

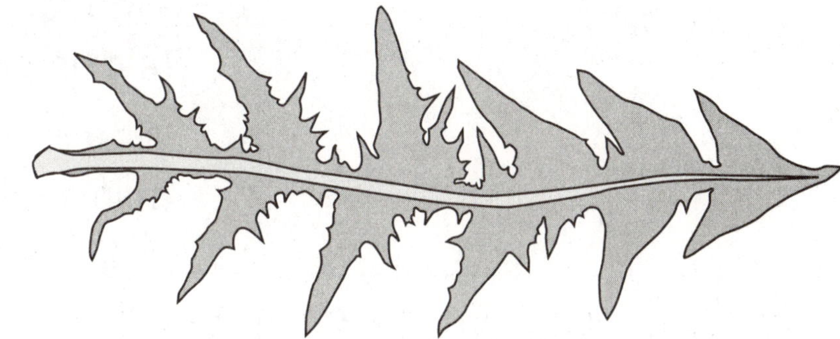

about ______ in.

about ______ cm

6. List the objects in order from shortest to longest.

____

____

____

____

____

Date Time

LESSON 7·7

Math Boxes

1. Draw hands to show 4:15.

2. **Books Read**

Number of Books	Juan	Lilly	Grace	Terell
	X X X X X	X X	X X X X X X X	X X X X

Who read the most books (maximum)? ________

Who read the least books (minimum)? ______

3. There are 6 rooms. Each room has 4 windows. How many windows in all? _____ windows

Draw an array.

MRB 112 113

4. Fill in the missing numbers.

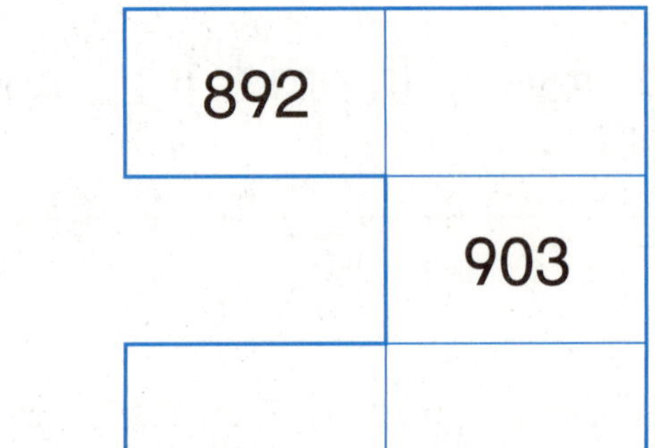

892	
	903

5. Here is a picture of a pyramid. What is the shape of one of the faces? _______________

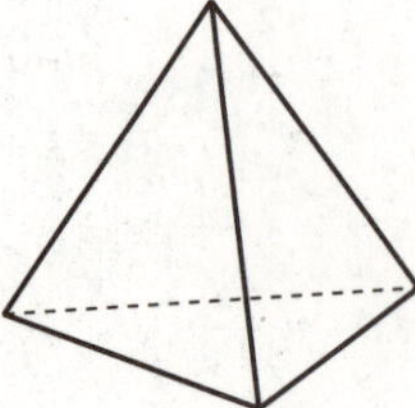

6. Color $\frac{5}{8}$ of the rectangle.

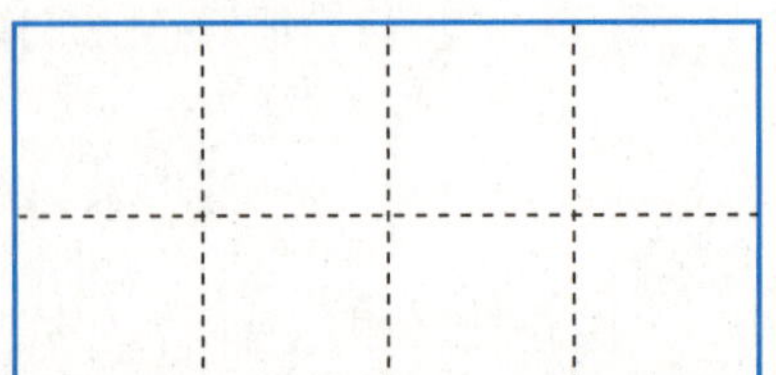

Date Time

Soccer Spin Directions

Materials ☐ *Math Masters,* pp. 470 and 471

☐ counter

☐ paper clip

☐ pencil

Players 2

Use a pencil and paper clip to make a spinner.

Skill Predict outcomes of events

Object of the Game To test the prediction made at the beginning of the game

Directions

1. Players agree upon one spinner to use during the game.
2. Each player chooses a team to cheer for, **Checks** or **Stripes.** (Players can cheer for the same team.) They look at their spinner choice and predict which team will win the game.
3. The game begins with the counter in the center of the soccer field.
4. Players take turns spinning and moving the counter one space toward the goal that comes up on the spinner.
5. The game is over when the counter reaches a goal.
6. Players compare and discuss the results of their predictions. Play two more games using the other two spinners.

Follow-Up

1. Which spinner(s) would you want to use if you were cheering for the **Checks** team? Explain.
2. Which spinner(s) would you want to use if you were cheering for the **Stripes** team? Explain.

Date Time

Table of Our Arm Spans

Make a table of the arm spans of your classmates.

Our Arm Spans

Arm Span (inches)	Frequency	
	Tallies	Number

Total =

Date Time

LESSON 7·8

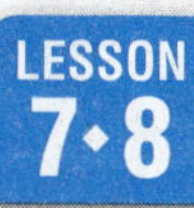

Bar Graph of Our Arm Spans

Make a bar graph of the arm spans of your classmates.

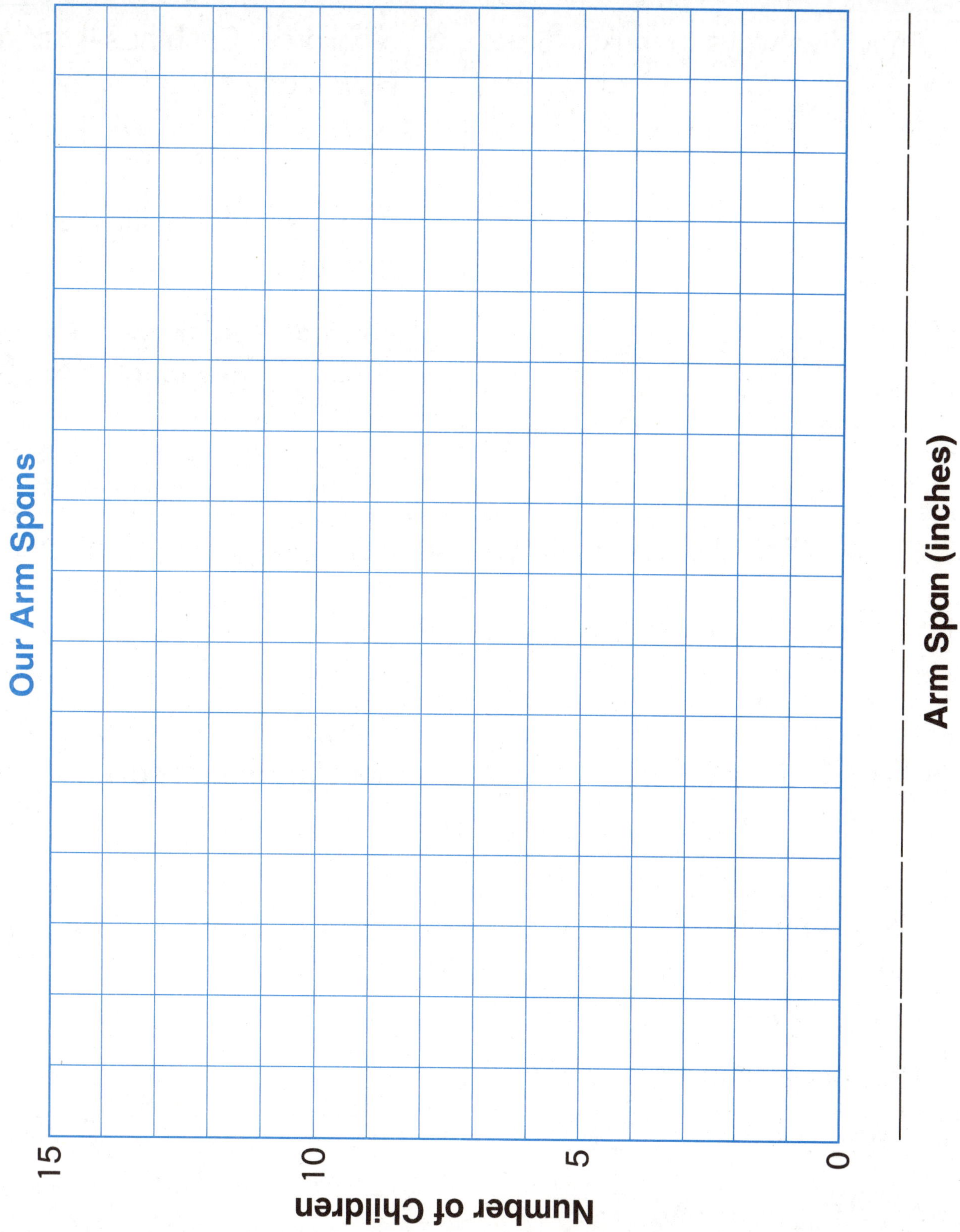
Our Arm Spans
Number of Children
0
5
10
15
Arm Span (inches)

Date Time

LESSON 7•8

Math Boxes

1. Show five ways to make 45¢.

2. 6 children. Each has 4 stickers. How many stickers in all?

_______ stickers

Complete the multiplication diagram.

children	stickers per child	stickers in all

3. Show 52° on the thermometer.

4. Put these numbers in order.
25, 15, 25, 19, 15, 75, 15

____

Which number is the most popular (the mode)?

____

5. Write a number model for a ballpark estimate. Solve.

Unit

Ballpark estimate:

____

$$\begin{array}{r} 47 \\ -\ 31 \\ \hline \end{array}$$

6. Write <, >, or = in the box.

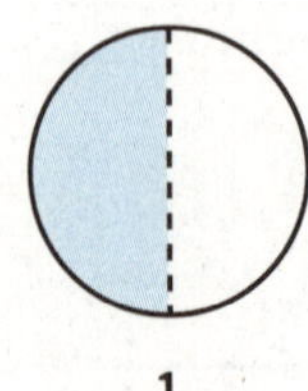

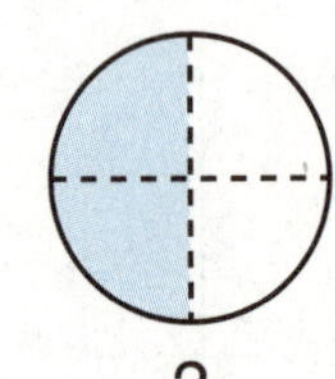

$\frac{1}{2}$  $\frac{2}{4}$

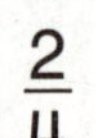

Date Time

LESSON 7•9 Math Boxes

1. What fraction of the triangles are shaded?

$\frac{\square}{5}$

2. Fill in the box.

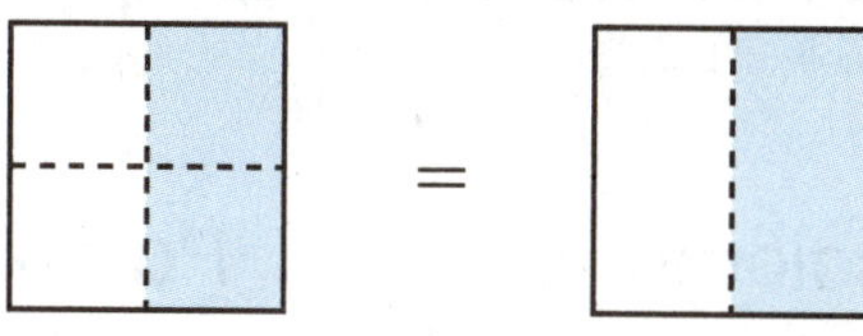

$\frac{\square}{4} = \frac{1}{2}$

3. Hana has 6 bracelets to wear. 3 of them are made of beads. What fraction of the bracelets are made of beads?

______

4. Shade $\frac{1}{3}$ of this trapezoid.

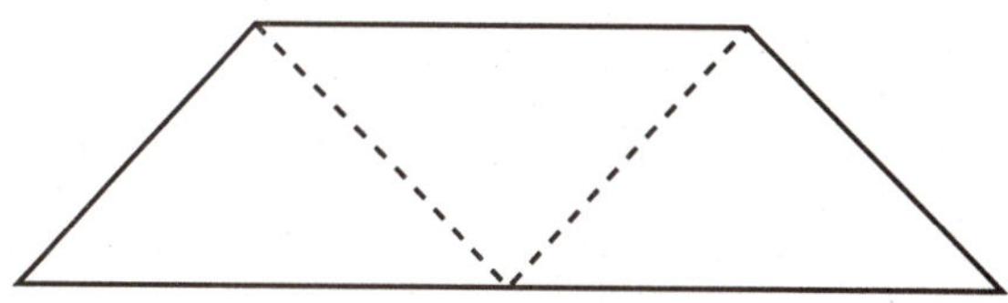

5. Circle $\frac{1}{4}$ of the pennies.

6. If [rhombus] = 1,

then [triangle] = ______

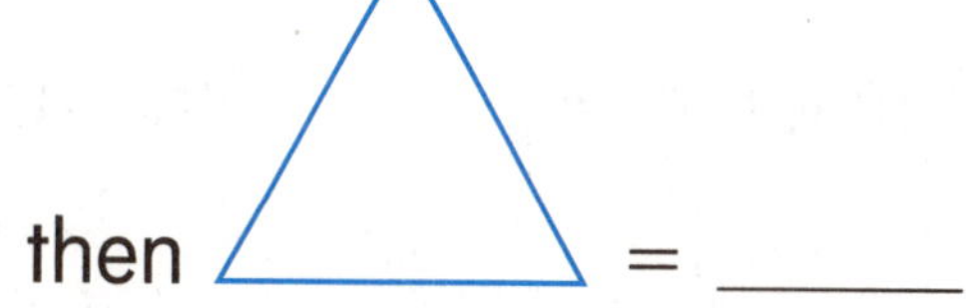

Date Time

Equal Parts

Use a straightedge or Pattern-Block Template.

1. Divide the shape into 2 equal parts. Color 1 part.

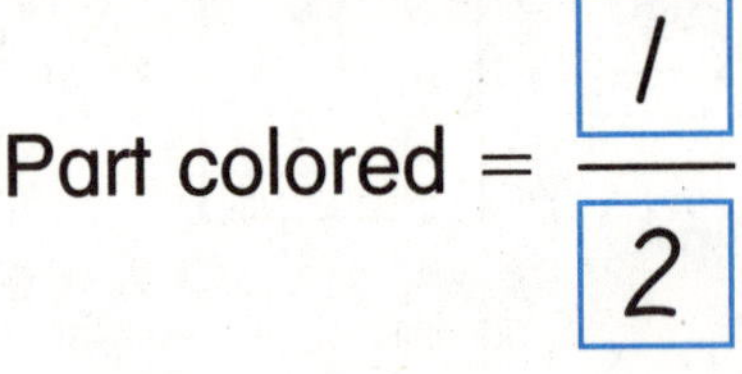

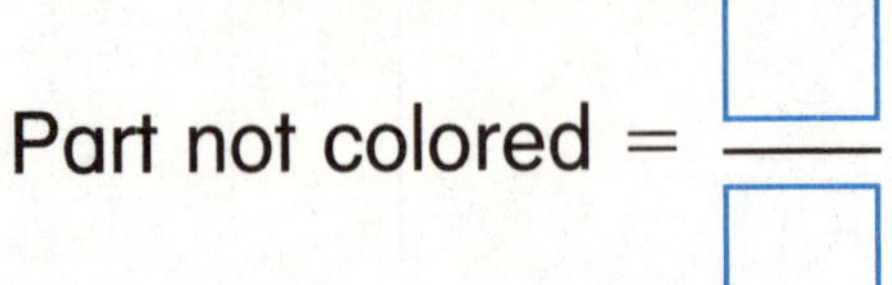

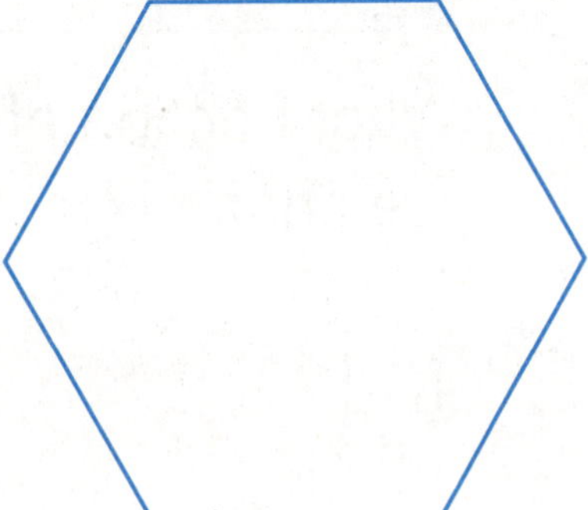

2. Divide the shape into 6 equal parts. Color 1 part.

Part colored = ☐/☐ Part not colored = ☐/☐

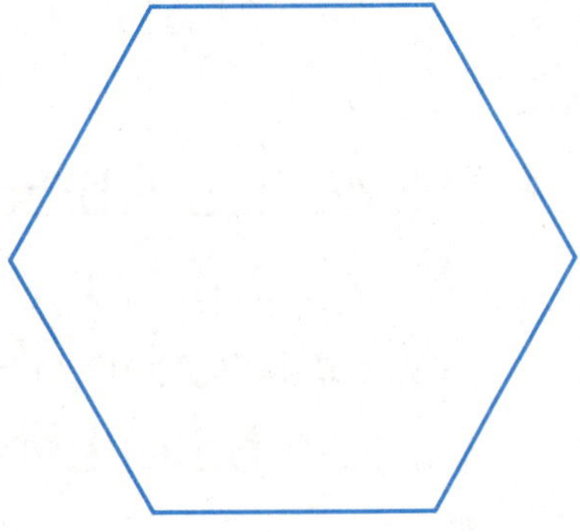

3. Divide the shape into 3 equal parts. Color 2 parts.

Part colored = ☐/☐ Part not colored = ☐/☐

4. Divide the shape into 4 equal parts. Color 2 parts.

Part colored = ☐/☐ Part not colored = ☐/☐

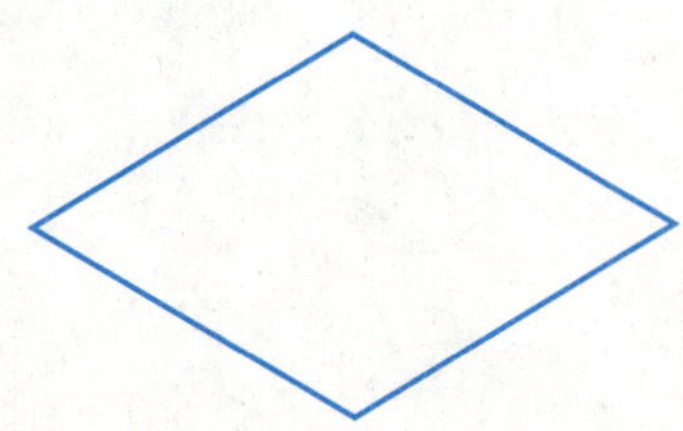

Date Time

Dressing for School

Jerome is deciding what to wear to school today. He has a red shirt, a blue shirt, and a green shirt. He has a black pair of pants, a yellow pair of pants, and an orange pair of pants. How many different outfits can Jerome make? Color the figures below to show the possible shirt and pants combinations that Jerome could wear.

How many possible outfits can Jerome make? ___________

Date Time

Math Boxes

1. Write fractions.

The part shaded = ____.

The part not shaded = ____.

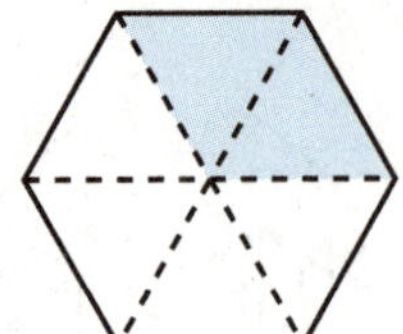

2. Use a Pattern-Block Template. Draw a shape that has at least one line of symmetry.

3. Complete the table.

in	out
2	1
4	
8	
	5

Rule

$\frac{1}{2}$ of

4. Show 2 ways to make 50¢. Use Ⓠ, Ⓓ, Ⓝ, and Ⓟ.

5. Circle the thing you are certain will happen.

You will roll a 7 on a die.

The temperature will be exactly 20°F today.

An hour will pass.

6.

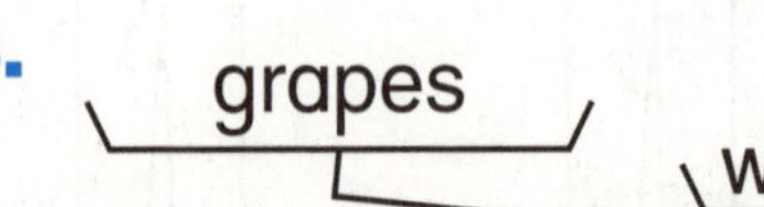

Which item is heavier?

____

Date Time

Pattern-Block Fractions

Use pattern blocks to help you solve each problem.

Use your Pattern-Block Template to show what you did.

Example:

If 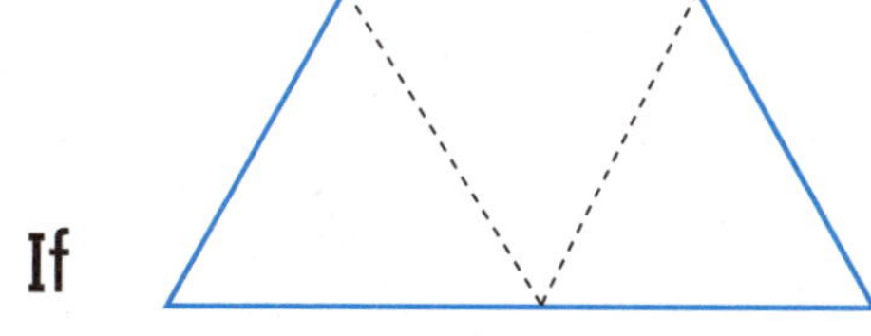 = 1, then 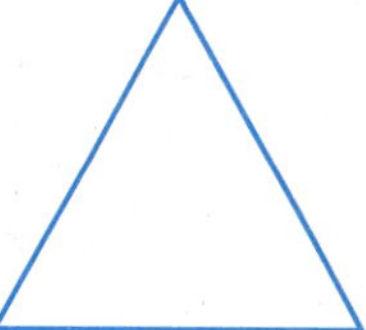 = $\frac{1}{3}$.

1. If = 1, then 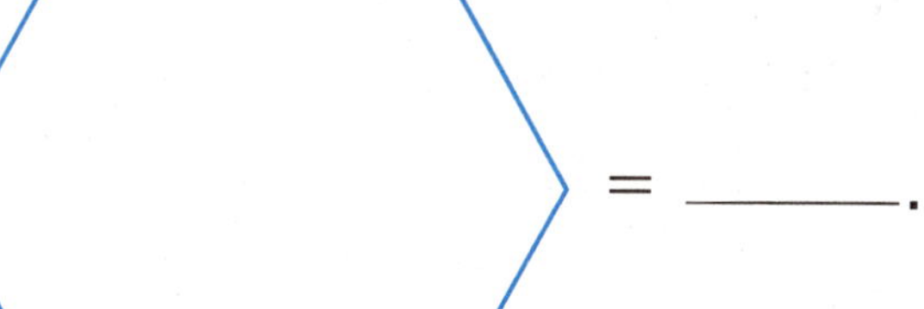 = ______.

2. If = 1, then  = ______.

Date Time

LESSON 8·2

Pattern-Block Fractions *continued*

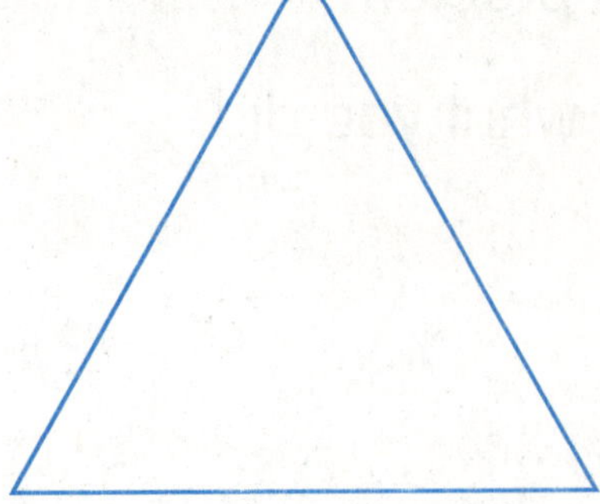

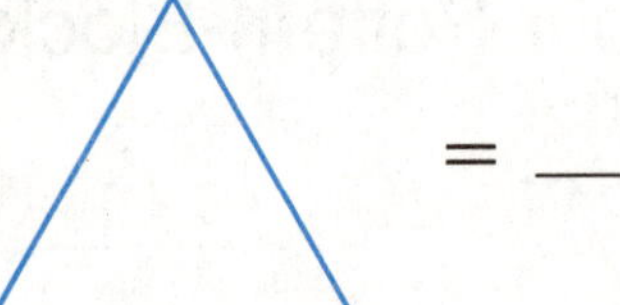

3. If = 1, then = ______.

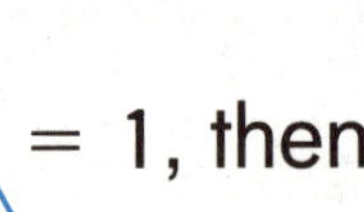

4. If = 1, then = ______.

5. If = 1, then = ______.

6. If = 1, then = ______.

Date Time

LESSON 8•2

Geoboard Fences

1.

2.

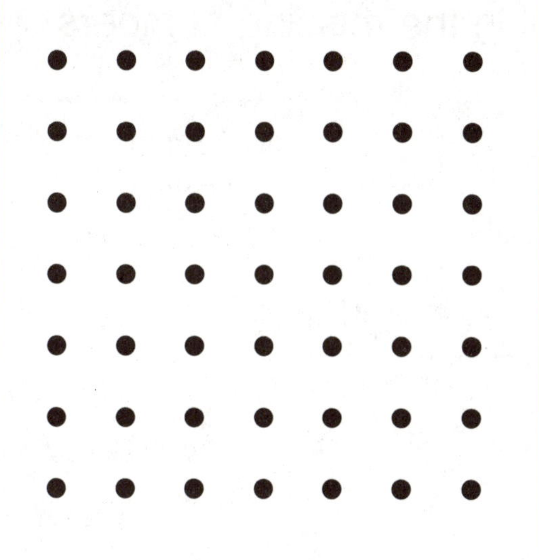

3.

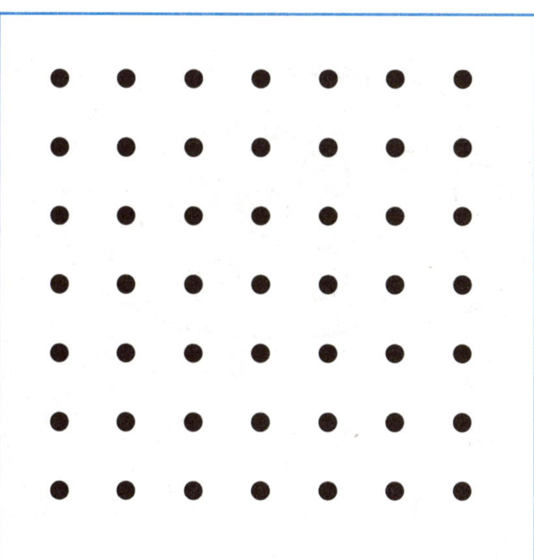

4.

Fence	How many pegs in all?	How many rows of pegs?	How many in each row?
1.			
2.			
3.			
4.			

LESSON 8·2

Math Boxes

1. Fill in the missing numbers.

189	
	200

156	
	167

2. 1 hour = ________ minutes

$\frac{1}{2}$ hour = ________ minutes

$\frac{1}{4}$ hour = ________ minutes

3. Put a line under the digit in the ones place.

479 364

1,796 5,079

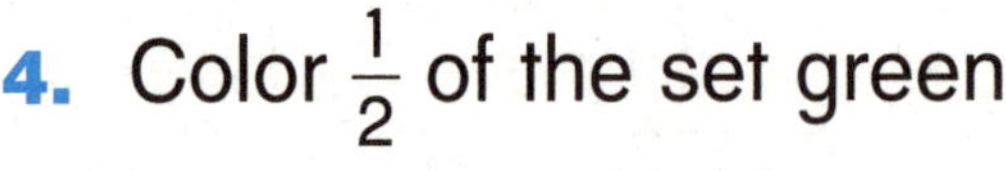

4. Color $\frac{1}{2}$ of the set green.

5. Measure the length of this line.

about ________ cm

about ________ in.

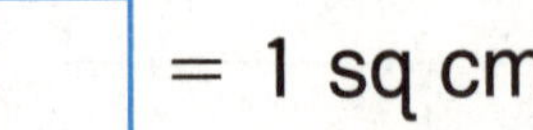

6. ☐ = 1 sq cm

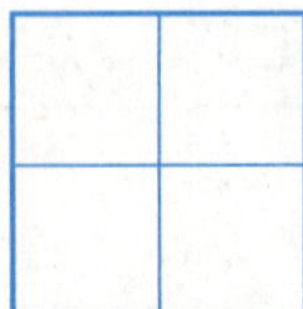

Area = ________ sq cm

MRB 69

Date Time

LESSON 8·3

Equal Shares

Use pennies to help you solve the problems.

Circle each person's share.

1. Two people share 10 pennies. How many pennies does each person get?

_____ pennies

$\frac{1}{2}$ of 10 pennies = _____ pennies.

2. Three people share 9 pennies. How many pennies does each person get?

_____ pennies

$\frac{1}{3}$ of 9 pennies = _____ pennies.

$\frac{2}{3}$ of 9 pennies = _____ pennies.

3. Four people share 12 pennies. How many pennies does each person get?

_____ pennies

$\frac{1}{4}$ of 12 pennies = _____ pennies.

$\frac{3}{4}$ of 12 pennies = _____ pennies.

Date Time

LESSON 8·3

Fractions of Sets

A fraction is given in each problem. Color that fraction of the checkers red.

1. $\frac{1}{5}$ are red.	**2.** $\frac{2}{3}$ are red.
3. $\frac{3}{4}$ are red.	**4.** $\frac{4}{6}$ are red.
5. $\frac{1}{2}$ are red.	**6.** $\frac{0}{7}$ are red.

Try This

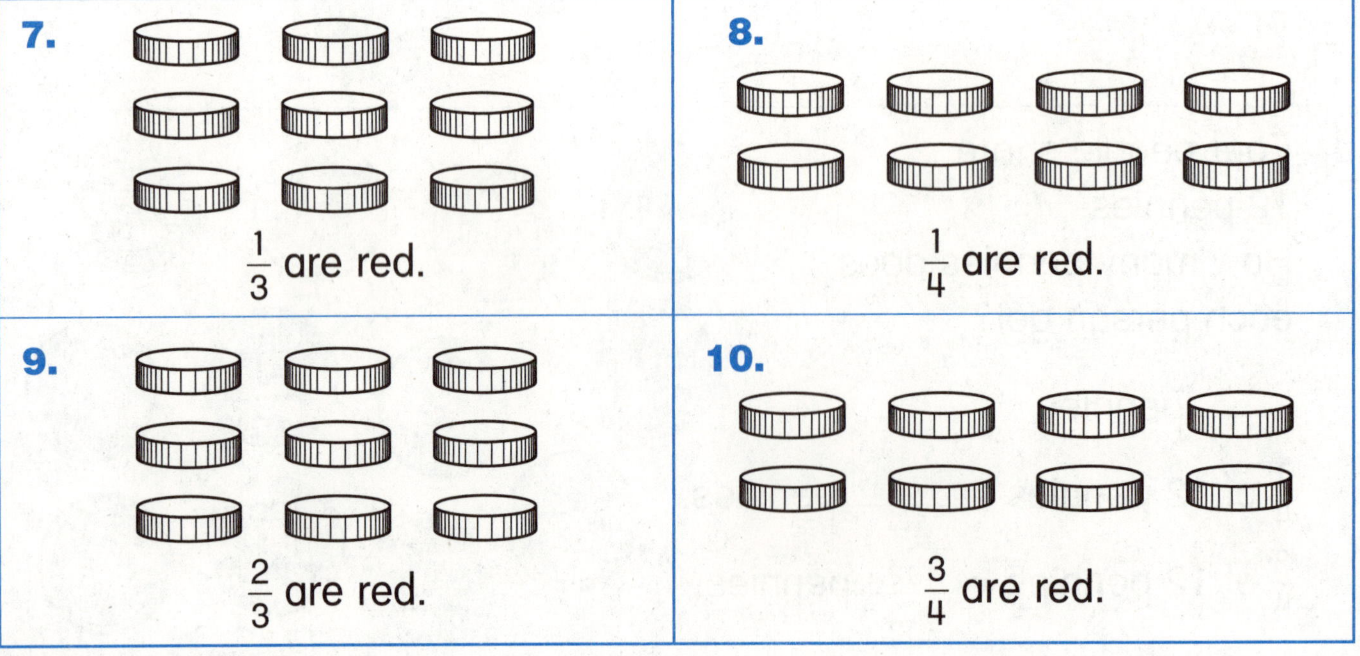

7. $\frac{1}{3}$ are red.	**8.** $\frac{1}{4}$ are red.
9. $\frac{2}{3}$ are red.	**10.** $\frac{3}{4}$ are red.

LESSON 8•3

Equal Parts

Use a straightedge or Pattern-Block Template.

1. Divide the shape into 2 equal parts. Color 1 part.

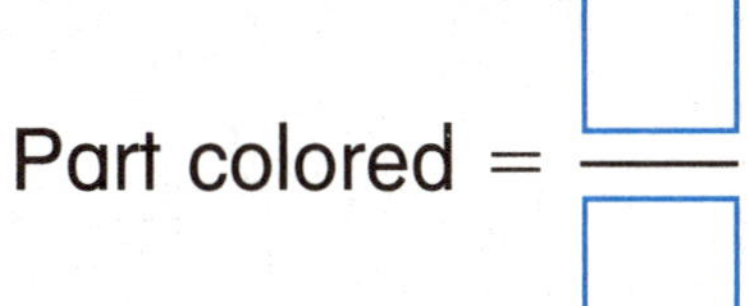

2. Divide the shape into 3 equal parts. Color 1 part.

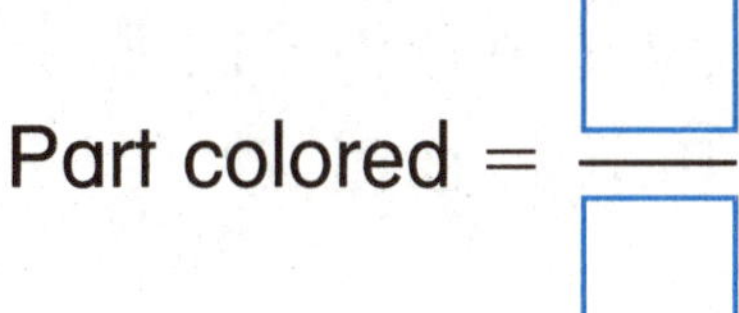

Part not colored =

3. Divide the shape into 4 equal parts. Color 2 parts.

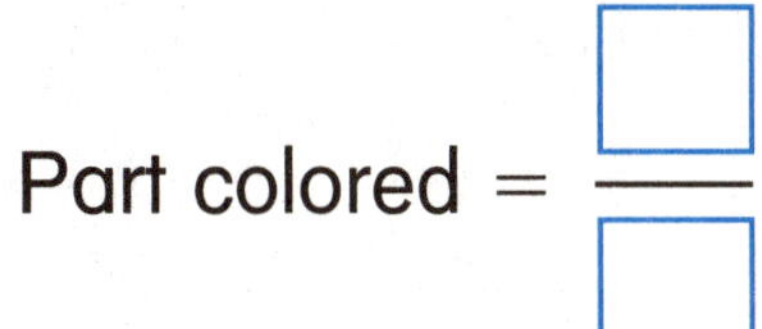

Part not colored =

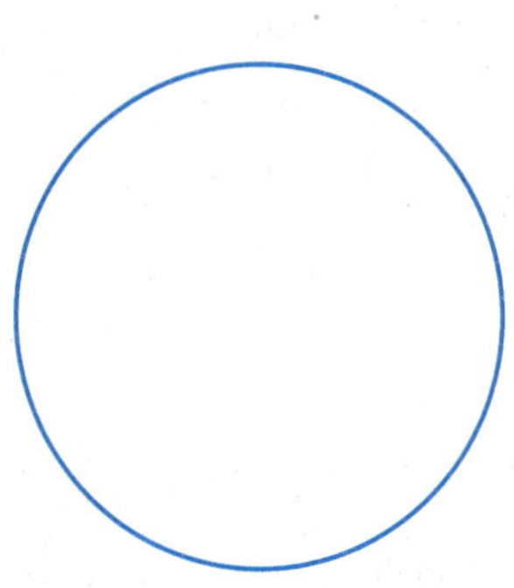

4. Divide the shape into 4 equal parts. Color 3 parts.

Part colored =

Part not colored =

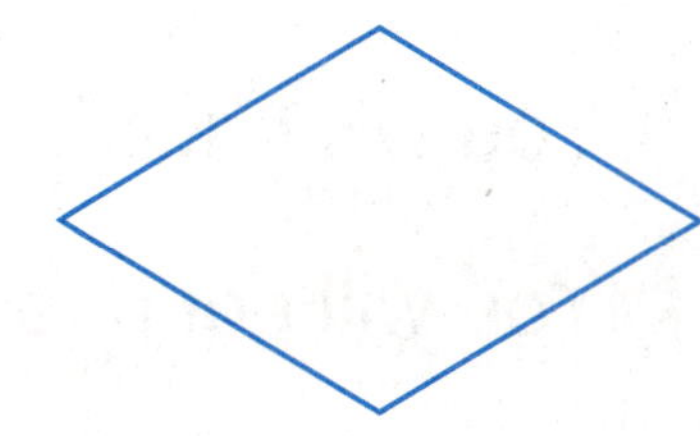

LESSON 8•3

Math Boxes

1. Color $\frac{1}{4}$ blue. Color $\frac{1}{4}$ yellow. Color $\frac{1}{2}$ red.

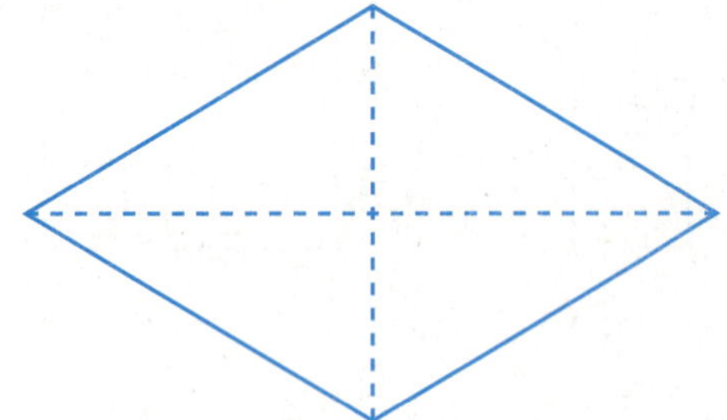

2. Circle the figure that has only one line of symmetry. Draw the line of symmetry.

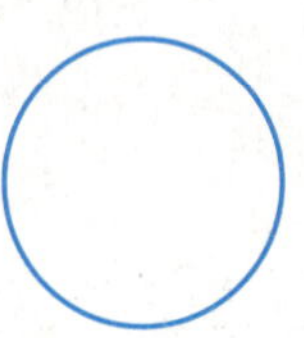 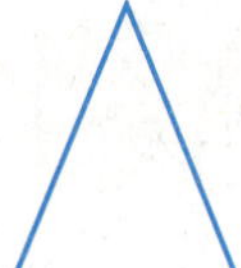 

3. Complete the table.

Rule
1 yd = 3 ft

yd	ft
2	
	9
5	
	30

4. Show 1 way to make $1.28. Use Ⓠ, Ⓓ, Ⓝ, and Ⓟ.

5. Circle the event that is likely to happen.

You will fly to the center of the earth.

You will have homework.

You will eat a rock.

6. Which unit makes sense? Choose the best answer.

A can of soup may weigh:

- ⬭ 8 ounces
- ⬭ 8 cups
- ⬭ 8 pounds
- ⬭ 8 feet

LESSON 8•4 Math Boxes

1. Fill in the missing numbers.

	1,217	

2. ________ months = 1 year

________ months = $\frac{1}{2}$ year

________ months = $\frac{1}{4}$ year

________ months = 2 years

3. Circle the digits in the hundreds place.

1 2 8 9 7 2 4 6 3

2, 4 6 5 3, 0 9 1

6 6, 2 5 0

4. There are 9 dinosaurs. 3 are plant eaters. Which fraction shows how many are plant eaters? Fill in the circle next to the best answer.

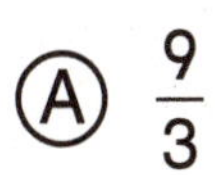
Ⓐ $\frac{9}{3}$

Ⓒ $\frac{1}{2}$

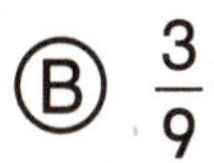
Ⓑ $\frac{3}{9}$

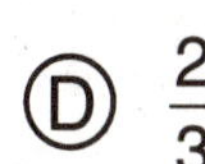
Ⓓ $\frac{2}{3}$

5. Draw a triangle. Measure each side to the nearest inch.

about ___ in.

about ___ in.

about ___ in.

6. Count the squares to find the area.

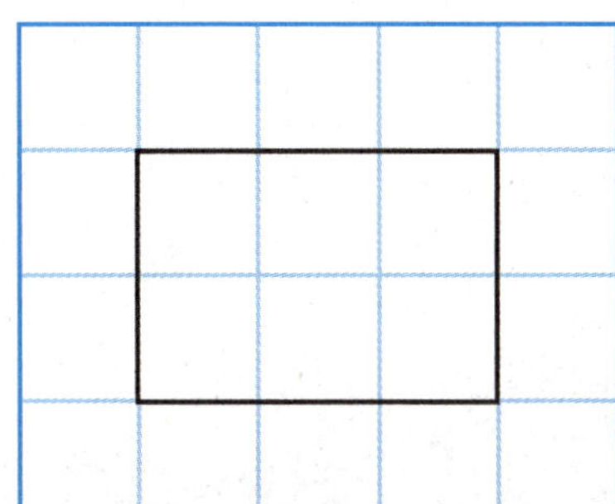

Area = ___ sq cm

LESSON 8•4

Equivalent Fractions

Do the following:

- Use the circles that you cut out of *Math Masters,* page 239.
- Cut these circles apart along the dashed lines.
- Glue the cutout pieces onto the circles on this page and the next, as directed.
- Write the missing numerators to complete the equivalent fractions.

1. Cover $\frac{1}{2}$ of the circle with fourths.

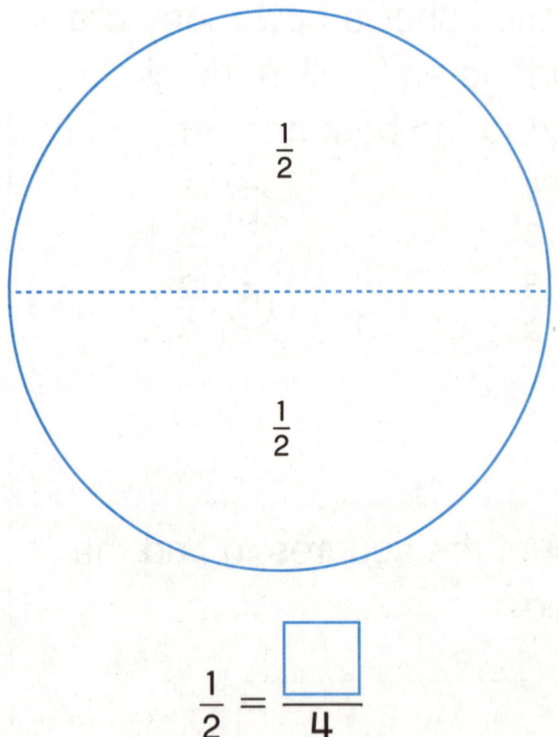

$\frac{1}{2} = \frac{\square}{4}$

2. Cover $\frac{1}{4}$ of the circle with eighths.

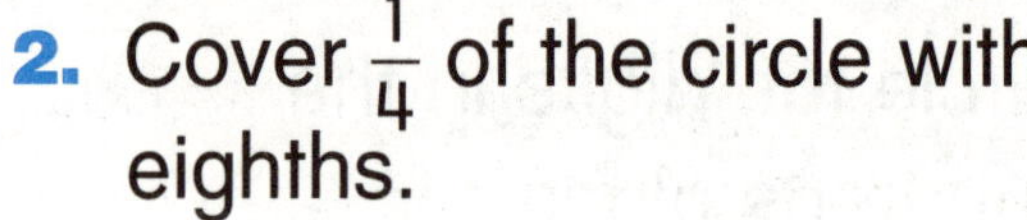

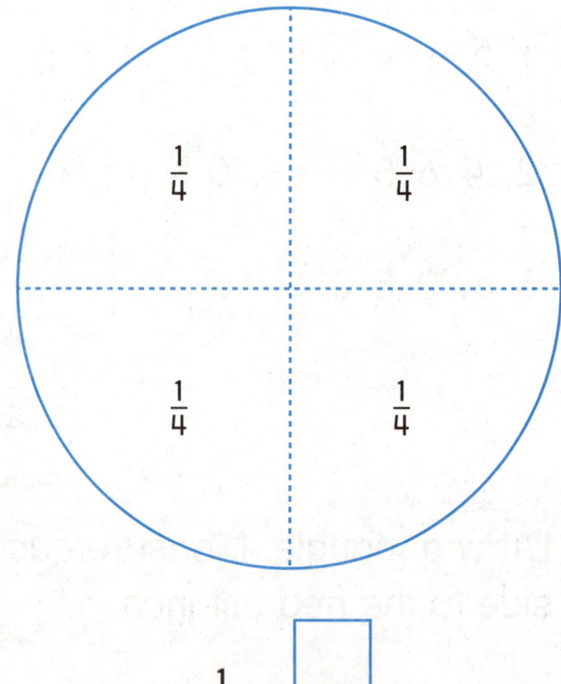

$\frac{1}{4} = \frac{\square}{8}$

LESSON 8·4

Equivalent Fractions *continued*

3. Cover $\frac{2}{4}$ of the circle with eighths.

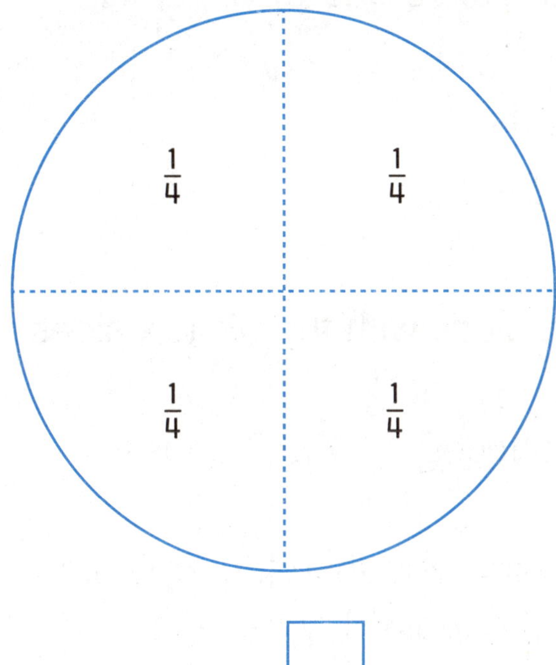

$\frac{2}{4} = \frac{\square}{8}$

4. Cover $\frac{1}{2}$ of the circle with sixths.

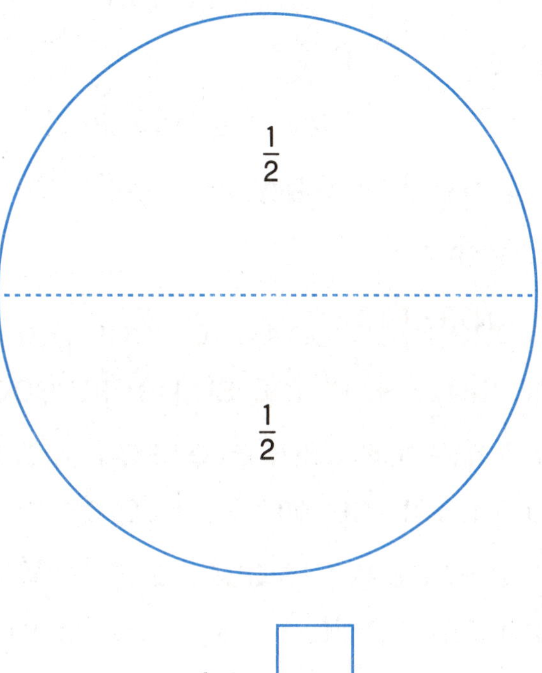

$\frac{1}{2} = \frac{\square}{6}$

5. Cover $\frac{1}{3}$ of the circle with sixths.

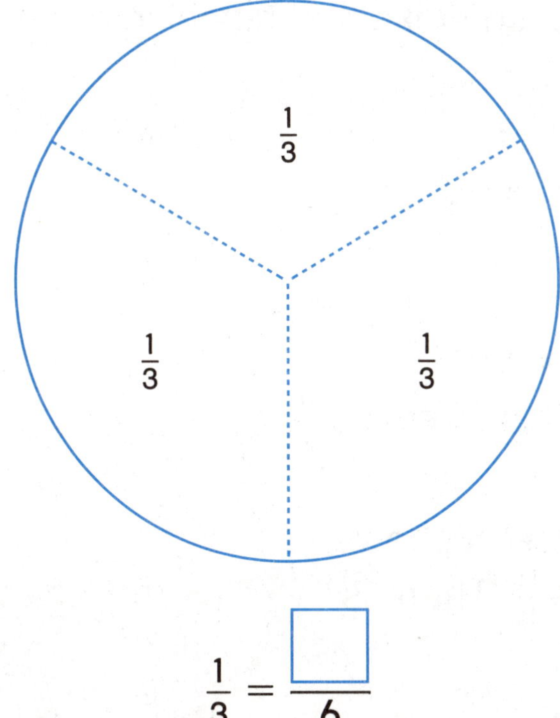

$\frac{1}{3} = \frac{\square}{6}$

6. Cover $\frac{2}{3}$ of the circle with sixths.

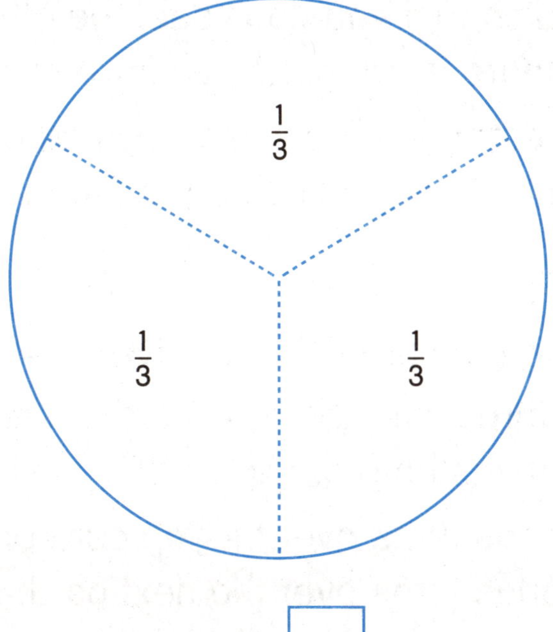

$\frac{2}{3} = \frac{\square}{6}$

LESSON 8•5

Equivalent Fractions Game Directions

Materials ☐ 32 Fraction Cards (2 sets cut from *Math Journal 2,* Activity Sheets 5 and 6)

Players 2

Skill Match equivalent fraction cards

Object of the Game To have the most cards

Directions

1. Mix the Fraction Cards and put them in a stack with the picture sides (the sides with the strips) facedown.
2. Turn the top card over so the picture side faces up. Put it on the table near the stack of cards.
3. Take turns with your partner. When it is your turn, take the top card from the stack. Turn it over and put it on the table. Try to match this card with a picture-side-up card on the table. (If there are no other picture-side-up cards on the table, turn over the next card on the stack and put it on the table.)
4. Look for a match. If two cards match, take both of them. If there is a match that you don't see, the other player can take the matching cards. If there is no match, your turn is over.
5. The game ends when each card has been matched with another card. The player who took more cards wins the game.

Example:

1. The top card is turned over. The picture shows $\frac{4}{6}$.
2. Li turns over the next card. It shows $\frac{2}{3}$. This matches $\frac{4}{6}$. Li takes both cards.
3. Carlos turns over the top card on the stack. It shows $\frac{6}{8}$. Carlos turns over the next card. It shows $\frac{0}{4}$. There is no match. Carlos places $\frac{0}{4}$ next to $\frac{6}{8}$.
4. Li turns over the top card on the stack.

Date Time

Equivalent Fractions Game Directions *continued*

Another Version

1. Mix the Fraction Cards and put them in a stack with the picture sides facedown.
2. Turn the top card over so the picture side faces up. Put it on the table with the picture side faceup.
3. Players take turns. When it is your turn, take the top card from the stack, but do *not* turn it over. Keep the picture side down. Try to match the fraction on the card with one of the picture-side-up cards on the table.
4. If you find a match, turn your card over. Check that your match is correct by comparing the two pictures. If your match is correct, take both cards.

   If there is no match, place your card next to the other cards, picture side faceup. Your turn is over. If the other player can find a match, he or she can take the matching cards.
5. If there are no picture cards showing when Player 2 begins his or her turn, take the top card from the stack. Place it on the table with the picture side showing. Then Player 2 takes the next card in the stack and doesn't turn that card over.

Marta thinks these two cards are a matching pair.

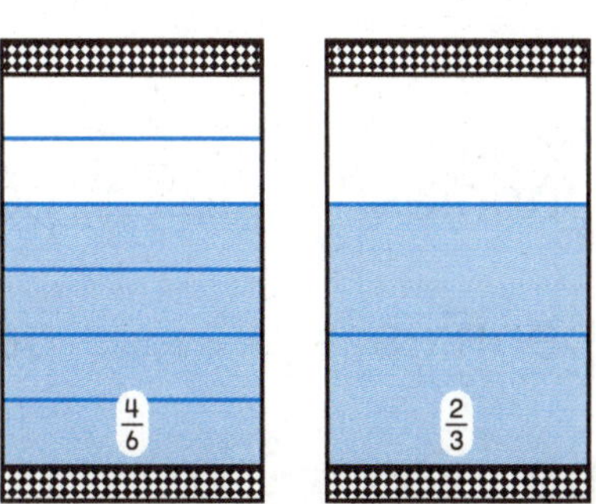

Marta checks that her match is correct by comparing the pictures of the fractions.

Date Time

LESSON 8•5

Fractions of Collections

Use pennies to help you solve the problems.

1. Five people share 15 pennies.

How many pennies does each person get? _____ pennies

$\frac{1}{5}$ of 15 pennies = _____ pennies.

$\frac{2}{5}$ of 15 pennies = _____ pennies.

2. Six people share 12 pennies.

How many pennies does each person get? _____ pennies

$\frac{1}{6}$ of 12 pennies = _____ pennies.

$\frac{4}{6}$ of 12 pennies = _____ pennies.

3. Four people share 16 pennies.

How many pennies does each person get? _____ pennies

$\frac{1}{4}$ of 16 pennies = _____ pennies.

$\frac{4}{4}$ of 16 pennies = _____ pennies.

$\frac{2}{4}$ of 16 pennies = _____ pennies.

$\frac{3}{4}$ of 16 pennies = _____ pennies.

$\frac{0}{4}$ of 16 pennies = _____ pennies.

Date Time

LESSON 8•5

Fractions of Collections *continued*

Color the fractions of circles blue.

4. $\frac{3}{5}$ are blue.

5. $\frac{1}{2}$ are blue.

6. $\frac{1}{3}$ are blue.

7. $\frac{2}{3}$ are blue.

8. $\frac{3}{5}$ are blue.

9. $\frac{3}{4}$ are blue.

Try This

10. $\frac{3}{8}$ are blue.

11. $\frac{2}{6}$ are blue.

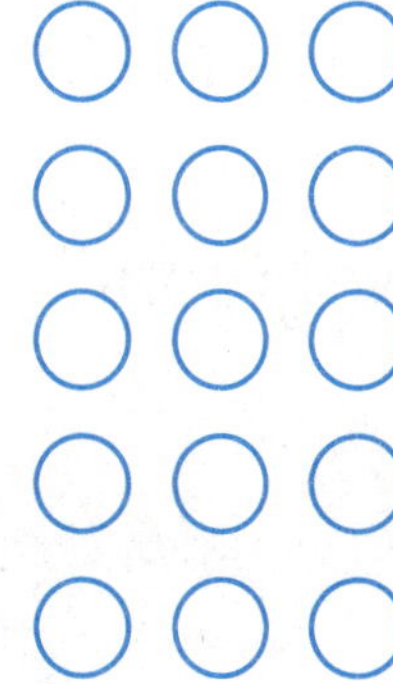

Date Time

Math Boxes

1. **Scores on a 5-Word Spelling Test**

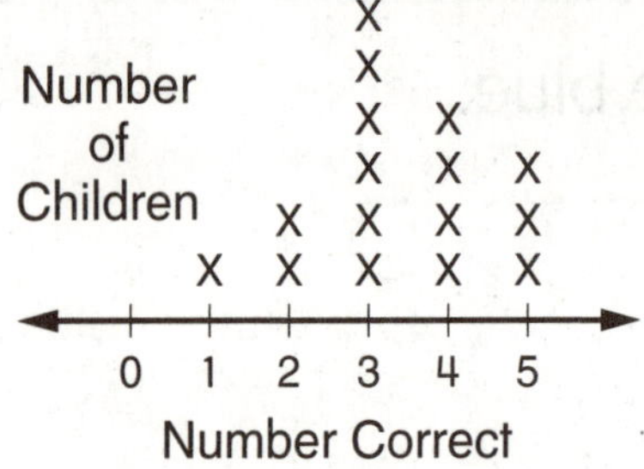

What score did the most children get (the *mode*)? _____

2. Circle $\frac{1}{5}$ of the nickels.

3. Find the arrow rules.

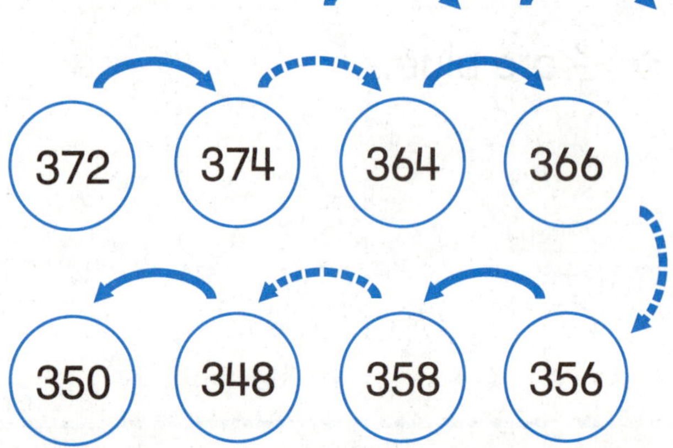

4. Complete the diagram.

Quantity 64	
Quantity 28	_____ Difference

Write a number model.

____

5. Draw a line segment $4\frac{1}{2}$ cm long.

Now draw a line segment 2 cm longer.

6. Measure each side of the triangle to the nearest inch.
Find the perimeter.

The perimeter is _____ inches.

LESSON 8•6

Fraction Top-It Directions

Use your Fraction Cards. List all the fractions that are:

less than $\frac{1}{2}$. ______________________

more than $\frac{1}{2}$. ______________________

the same as $\frac{1}{2}$. ______________________

Fraction Top-It

Materials ☐ 32 Fraction Cards (2 sets cut from *Math Journal 2,* Activity Sheets 5 and 6)

Players 2

Skill Compare fractions

Object of the Game To have the most cards

Directions

1. Mix the Fraction Cards and put them in a stack so all the picture sides (the sides with the strips) are facedown.
2. Each player turns over a card from the top of the stack. Players compare the shaded parts of their cards. The player with the larger (higher) fraction takes both cards.
3. If the shaded parts are equal, the fractions are equivalent. Each player turns over another card. The player with the larger fraction takes all the cards from both plays.
4. The game ends when all the cards have been taken from the stack. The player who took more cards wins.

$\frac{1}{2}$ $\frac{1}{2}$ is greater than $\frac{1}{3}$. $\frac{1}{3}$

Fraction Top-It Directions *continued*

Another Version

1. Mix the Fraction Cards and put them in a stack so all the picture sides (the sides with the strips) are facedown.

2. Each player takes a card from the top of the stack but does *not* turn it over.

3. Players take turns. When it is your turn, compare the fractions on the two cards. Say one of the following:

   - My fraction is more than your fraction.
   - My fraction is less than your fraction.
   - The fractions are equivalent.

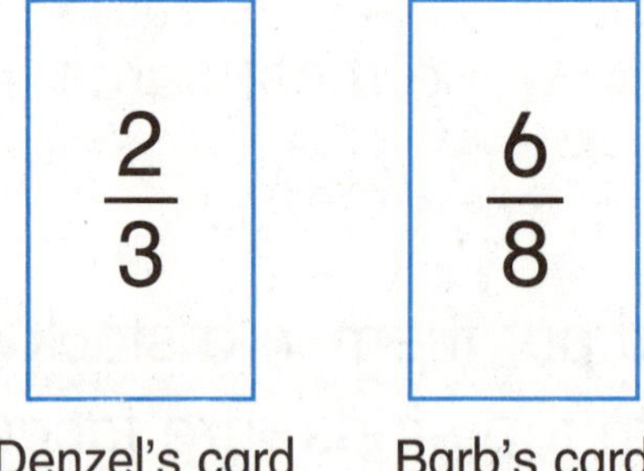

Denzel says that his fraction is less than Barb's fraction.

4. Turn the cards over and compare the shaded parts. If you were correct, take both cards. If you were not correct, the other player takes both cards.

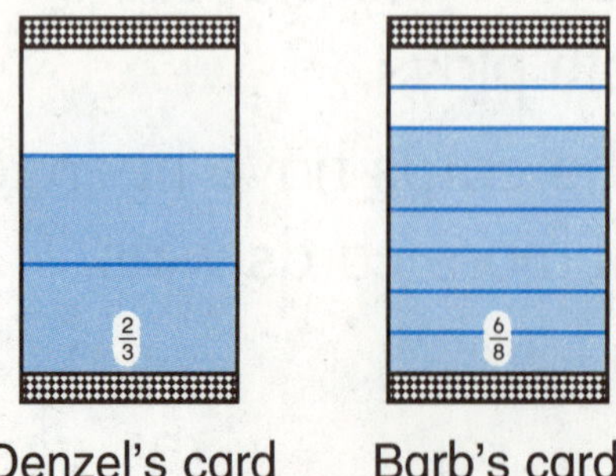

Less of Denzel's card is shaded: $\frac{2}{3}$ is less than $\frac{6}{8}$. Denzel takes both cards.

Date Time

Math Boxes

1. Fill in the missing numbers.

992	
	1,003

2. There are

_____ minutes in an hour.

_____ hours in a day.

_____ days in a week.

_____ weeks in a year.

3. 368

The value of 3 is _____.

The value of 6 is _____.

The value of 8 is _____.

MRB 10

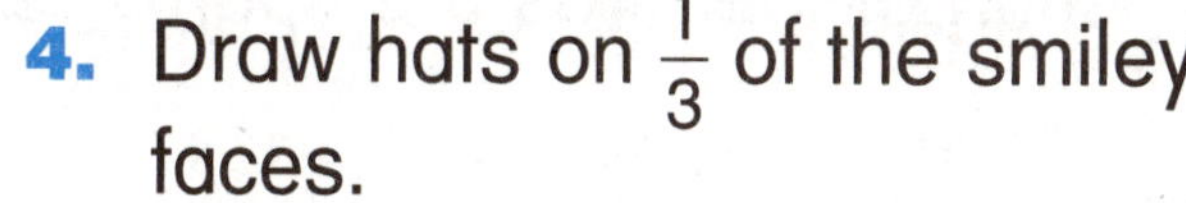

4. Draw hats on $\frac{1}{3}$ of the smiley faces.

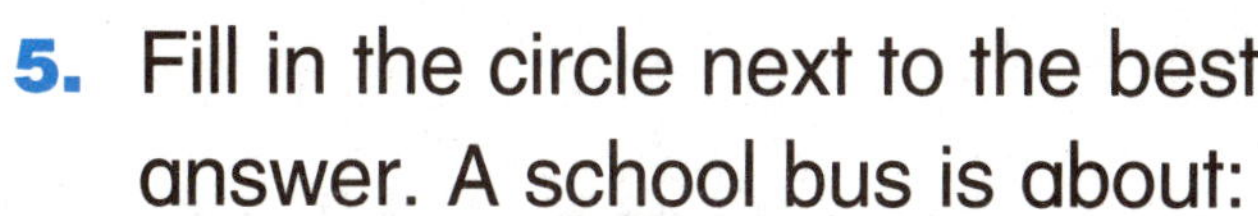

5. Fill in the circle next to the best answer. A school bus is about:

Ⓐ 180 cm long.

Ⓑ 18 m long.

Ⓒ 18 in long.

Ⓓ 180 m long.

6. ☐ = 1 sq cm

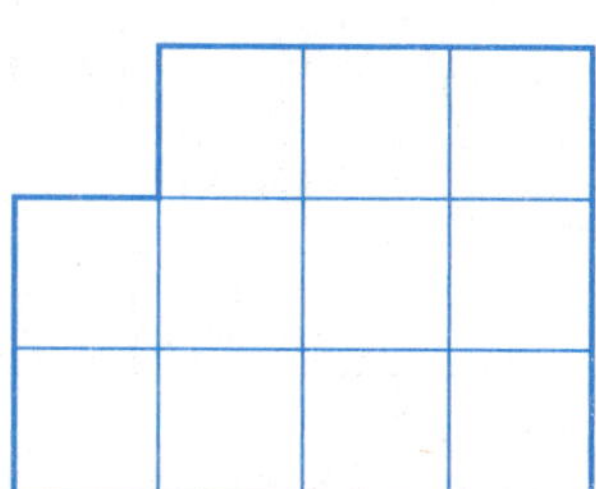

Area = _____ sq cm

Date Time

LESSON 8·7

Fraction Number Stories

Solve these number stories. To help, you can use pennies or other counters, or you can draw pictures.

1. Mark has 4 shirts to wear. 3 of them have short sleeves. What fraction of the shirts have short sleeves?

____

2. 8 birds are sitting on a tree branch. 6 of the birds are sparrows. What fraction of the birds are sparrows?

____

3. June has 15 fish in her fish tank. $\frac{1}{3}$ of the fish are guppies. How many guppies does she have?

____

Try This

4. Sam ate $\frac{0}{5}$ of a candy bar. How much of the candy bar did he eat?

____

5. If you were thirsty, would you rather have $\frac{2}{2}$ of a carton of milk or $\frac{4}{4}$ of that same carton? Explain.

____

____

____

____

Date Time

LESSON 8•7

Math Boxes

1. **Baskets Made by 2nd Graders**

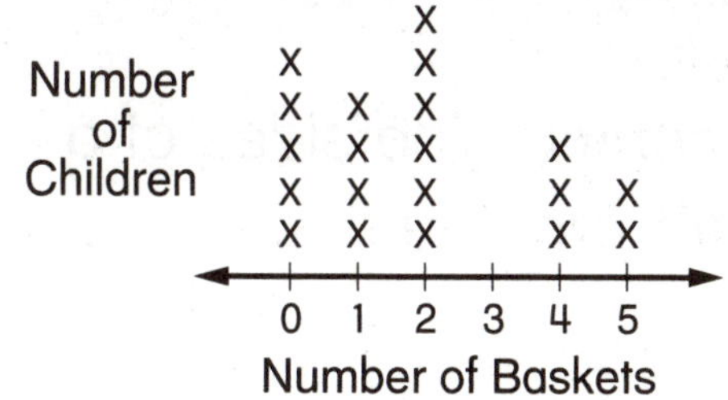

What was the most frequent number of baskets made (the *mode*)? ______

2. What fraction of dots is circled? Circle the best answer.

A $\frac{1}{2}$

B $\frac{1}{4}$

C $\frac{1}{3}$

D $\frac{2}{4}$

3. Fill in the missing numbers.

−5

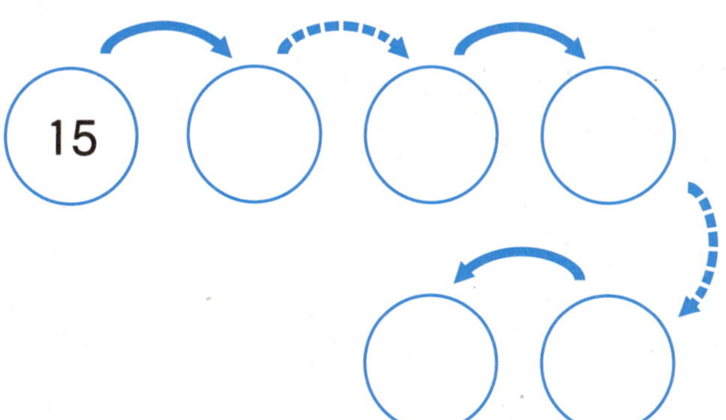

4. Complete the diagram. Then write a number model.

Quantity
82

Quantity	
39	______ Difference

____

5. Draw a line segment that is 6 cm long. Divide the line segment into 3 equal parts.

Each part = ____ cm

6. Measure each side to the nearest cm. Find the perimeter.

The perimeter is ____ cm.

MRB 68

Date Time

LESSON 8•8

Math Boxes

1. Circle. How much does a can of soda hold?

1 ounce

1 gallon

36 liters

12 ounces

2. Draw a square with a perimeter of 8 cm.

Remember: The sides of a square are equal.

3. Draw a rectangle. Measure each side to the nearest cm.

about _____ cm

about _____ cm

about _____ cm

about _____ cm

4. Measure the length of this line.

____

about _____ cm

about _____ in.

5. Match the items to the weights.

1 cat	1 ounce
3 envelopes	1 pound
1 book	7 pounds

6. ☐ = 1 sq cm

Area = _____ sq cm

Date Time

Yards

Materials ☐ yardstick

Directions

Record each step in the table below.

1. Choose a distance.
2. Estimate the distance in yards.
3. Use a yardstick to measure the distance to the nearest yard. Compare this measurement to your estimate.

Distance I Estimated and Measured	My Estimate	My Yardstick Measurement
	about _____ yards	about _____ yards
	about _____ yards	about _____ yards
	about _____ yards	about _____ yards
	about _____ yards	about _____ yards
	about _____ yards	about _____ yards

Date Time

Possible Outcomes

Solve the problem.

Your teacher placed 5 buttons in a bag. One button is a white square, 2 buttons are round and black, and 2 buttons are round and white.

Your teacher wants you to pull two buttons out of the bag while keeping your eyes shut. Draw a picture of all the button combinations that you could pull out of the bag.

How many different ways are there to pull two buttons out of the bag?

_____

Date Time

Math Boxes

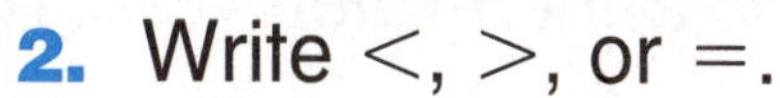

1. Make a square array with 25 pennies. How many pennies are in each row?

____ pennies

2. Write $<$, $>$, or $=$.

4 + 5 + 6 ______ 3 + 5 + 7

7 + 5 + 9 ______ 6 + 6 + 8

2 + 11 + 4 ______ 7 + 1 + 9

15 + 7 + 5 ______ 9 + 9 + 9

3. Circle the parallel lines.

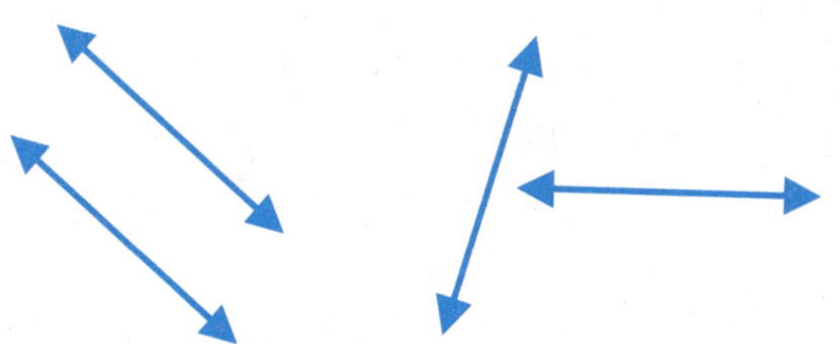

4.

Unit
yards

246 228 273
209 298

The median number of yards is ______.

5. Use your Pattern-Block Template. Trace a shape and draw 1 line of symmetry.

MRB 60

6. Count by quarters to \$3.00.

\$0.50, ______, ______,

______, ______, ______,

______, ______, ______,

______, ______

Date Time

LESSON 9•2 Units of Linear Measure

Materials ☐ 12-inch ruler ☐ 10-centimeter ruler

Directions

1. Measure the length of two objects or distances.
2. First measure to the nearest foot. Measure again to the nearest inch.
3. Then measure to the nearest decimeter. Measure again to the nearest centimeter.

Object *or* Distance	Nearest Foot	Nearest Inch
	about _____ ft	about _____ in.
	about _____ ft	about _____ in.

Object *or* Distance	Nearest Decimeter	Nearest Centimeter
	about _____ dm	about _____ cm
	about _____ dm	about _____ cm

"What's My Rule?"

4.

Rule
1 ft = 12 in.

ft	in.
1	
2	
	36

5.

Rule
1 m = 100 cm

m	cm
1	
	300
10	

Date Time

LESSON 9•2

Math Boxes

1. Fill in the missing numbers.

			1,250
		1,259	
	1,268		

2. Draw a line segment that is 8 cm long. Now draw a line segment 5 cm shorter.

3. Dillon leaped 32 inches. Marcus leaped 27 inches. How many more inches did Dillon leap? _____ inches

Fill in the diagram.

Quantity

Quantity

_____ **Difference**

4. What is the chance that you will have two birthdays this year? Choose the best answer.

- ⬭ unlikely
- ⬭ likely
- ⬭ certain
- ⬭ impossible

5. 3 insects. 6 legs per insect. How many legs in all?

_____ legs

Fill in the diagram and write a number model.

insects	legs per insect	legs in all

_____ × _____ = _____

6. Solve.

_____ pennies = \$2.00

_____ nickels = \$2.00

_____ dimes = \$2.00

_____ quarters = \$2.00

Date _______________ Time _______________

LESSON 9•3

Measuring Lengths with a Ruler

Materials ☐ inch ruler

☐ centimeter ruler

Directions

Work with a partner. Use your ruler to measure the length of each object to the nearest inch and centimeter.

1. large paper clip 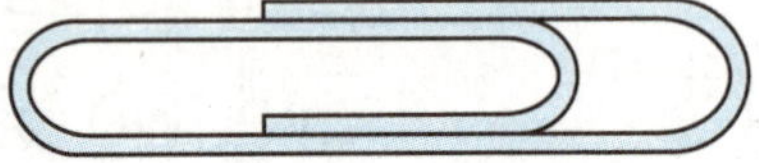

about _____ inches long about _____ centimeters long

2. pencil 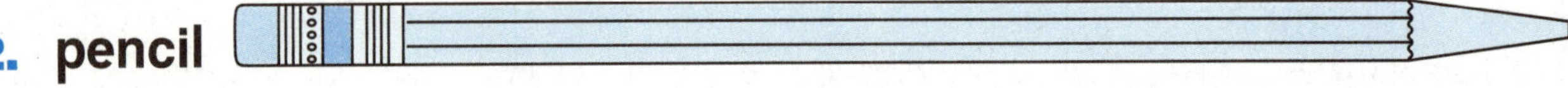

about _____ inches long about _____ centimeters long

3. nail 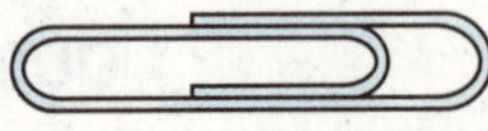

about _____ inch long about _____ centimeters long

Try This

Measure to the nearest $\frac{1}{2}$-inch and $\frac{1}{2}$-centimeter.

4. small paper clip

about _____ inches long about _____ centimeters long

Math Boxes

1. Make a square array with 36 pennies. How many pennies in each row?

____ pennies

2. Complete each number model.

Unit

________________ > 199

$372 >$ ________________

________________ $< 2{,}424$

$5{,}269 <$ ________________

3. Draw a quadrangle. Make 2 sides parallel.

MRB 51 55

4. 264 246 310
277 301

Unit
meters

Find the median number of meters. Circle the best answer.

A. 264 **B.** 277

C. 301 **D.** 310

5. Find 2 lines of symmetry.

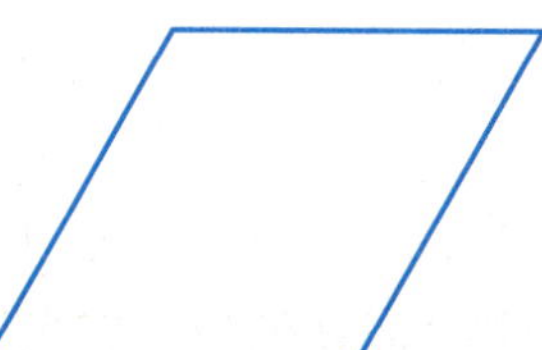

6. How much in all?

____

(D)(D)
(Q)(Q)(Q)(N)

[$20] [$5]
[$10] [$5]

LESSON 9•4 Distance Around and Perimeter

Measure the distance around the following to the nearest centimeter.

1. Your neck: ______ cm
2. Your ankle: ______ cm

Measure the distance around two other objects to the nearest centimeter.

3. Object: ______________________ Measurement: ______ cm
4. Object: ______________________ Measurement: ______ cm

Measure each side of the figure to the nearest inch. Write the length next to each side. Then find the perimeter.

5.

Perimeter: ______ inches

Try This

6.

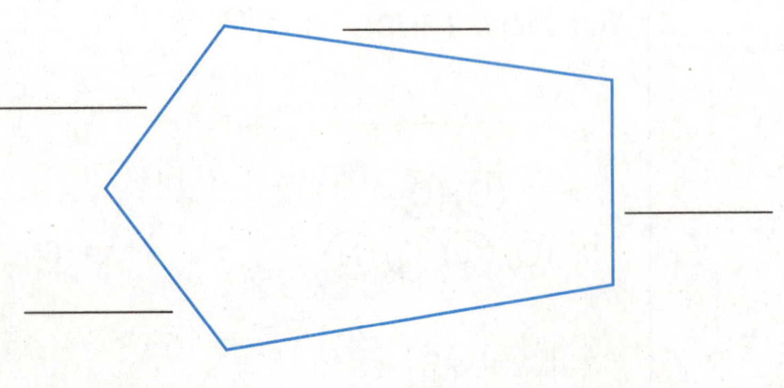

Perimeter: ______ inches

Date Time

LESSON 9·4

Math Boxes

1. Fill in the missing numbers.

	1,789	
	1,799	

2. Draw a line segment that is $3\frac{1}{2}$ inches long.

Now draw a line segment that is 1 inch shorter.

3. The Jays scored 63 points. The Gulls scored 46 points. How many more points did the Jays score? _____ points

Fill in the diagram.

Quantity

Quantity

____
Difference

4. What is the chance that you will fly in a spaceship today? Circle your answer.

impossible

certain

likely

unlikely

5. 9 cars. Each has 4 tires. How many tires in all?

cars	tires per car	tires in all

6. Use Ⓟ, Ⓝ, Ⓓ, and Ⓠ. Show $1.79.

Date Time

LESSON 9•5

Driving in the West

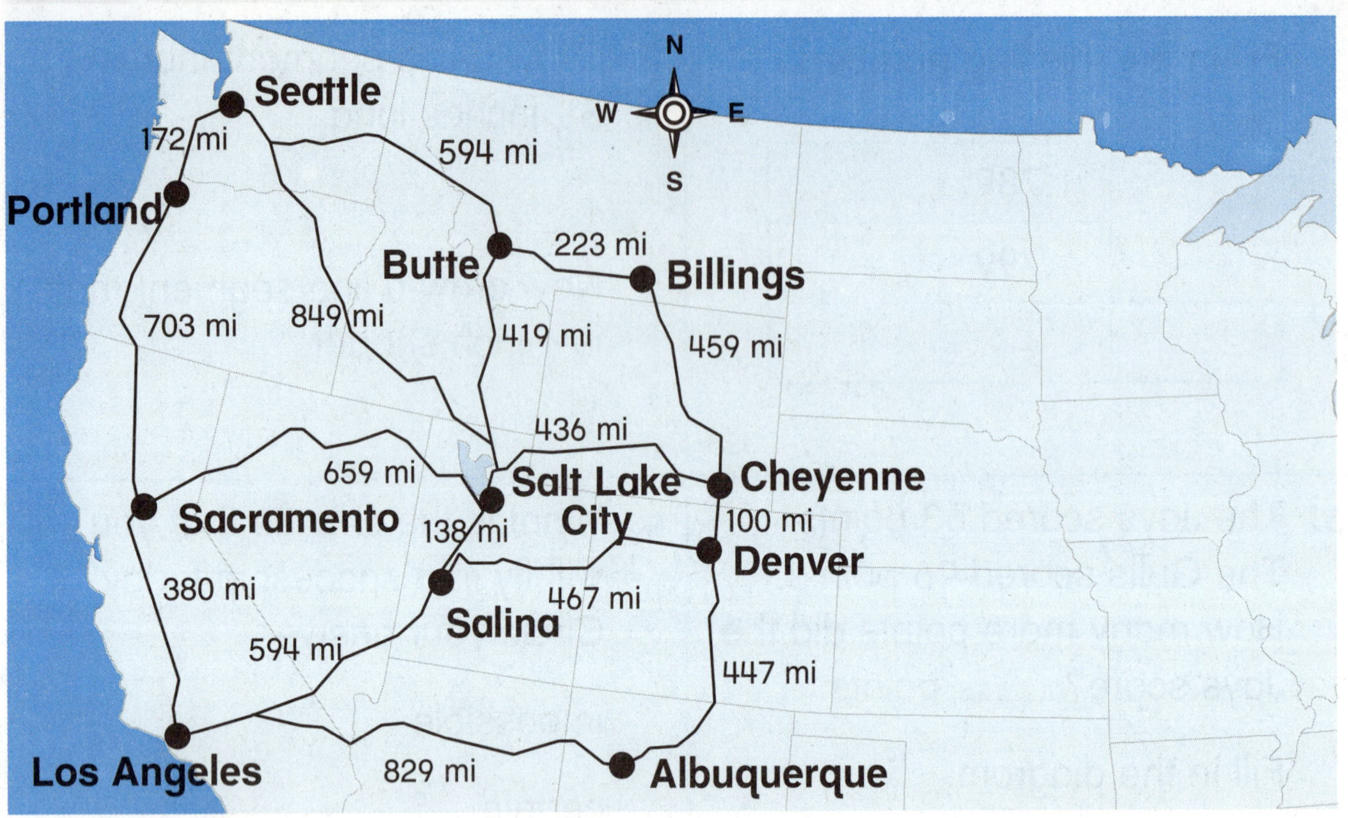

1. What is the shortest route from Seattle to Albuquerque?

____

____

2. Put a check mark in front of the longer trip.

______ Salt Lake City to Billings by way of Butte

______ Salt Lake City to Billings by way of Cheyenne

How much longer is that trip? about ______ miles longer

Date Time

LESSON 9•5

Addition and Subtraction Practice

Make a ballpark estimate. Solve. Compare your answer to your estimate.

1. Ballpark estimate:

____

45 + 68 = _______

2. Ballpark estimate:

____

143 + 78 = _______

3. Ballpark estimate:

____

158 + 233 = _______

4. Ballpark estimate:

____

74 − 49 = _______

5. Ballpark estimate:

____

133 − 86 = _______

6. Ballpark estimate:

____

256 − 147 = _______

LESSON 9•5

Math Boxes

1. Write 5 names for $\frac{1}{2}$. Use your Fraction Cards if you need help.

$\frac{1}{2}$

2. Write even or odd.

126 ________

311 ________

109 ________

430 ________

3. 17 pieces of gum are shared equally. Each child gets 3 pieces.

How many children are sharing?

______ children

How many pieces are left over?

______ pieces

4. Rosita had $0.39 and found $0.57 more. How much does she have now? Estimate your answer and then use partial sums to solve.

Estimate:

________ + ________ = ________

Answer: ________

5. In Pensacola, Florida, the temperature is 82°F. In Portland, Maine, the temperature is 64°F. What is the difference? Fill in the circle next to the best answer.

Ⓐ 22°F Ⓑ 17°F

Ⓒ 20°F Ⓓ 18°F

6. A number has:

7 thousands
8 tens
5 ten-thousands
1 one
0 hundreds

Write the number. ________

Date Time

Math Message

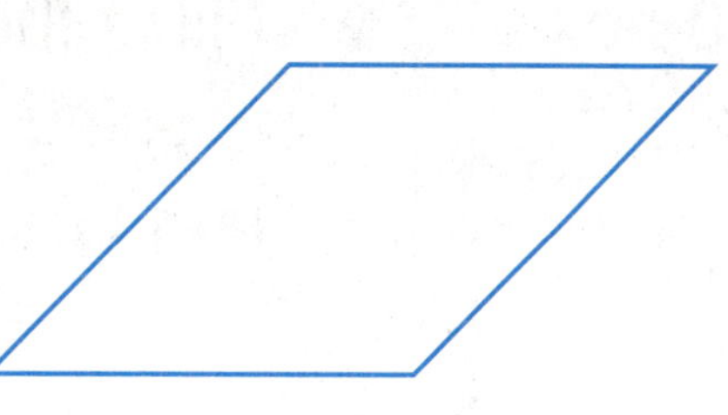

Estimate: Which shape is the "biggest" (has the largest area)? Circle it.

Think: How might you measure the shapes to find out?

Exploration A: Which Cylinder Holds More?

Which holds more macaroni—the tall and narrow cylinder or the short and wide cylinder?

My prediction: ______________________________

Actual result: ______________________________

Exploration B: Measuring Area

The area of my tracing of the deck of cards is about _____ square centimeters.

The area of my tracing of the deck of cards is about _____ square inches.

I traced ______________________________.

It has an area of about ______________________________.
(unit)

Date Time

LESSON 9•6

Math Boxes

1. Write 5 names in the 90-box.

MRB 16

2. Fill in the missing numbers.

MRB 98 99

3. Solve. Show your work.

$$\begin{array}{r} 27 \\ +\ 56 \\ \hline \end{array}$$

4.

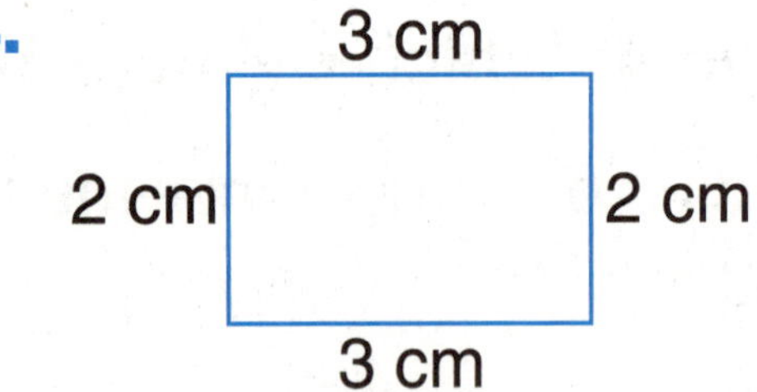

Perimeter = ______ cm

5. Write a number model for a ballpark estimate. Then solve.

Ballpark estimate:

____

$$\begin{array}{r} 68 \\ +\ 34 \\ \hline \end{array}$$

6. The total cost is 60¢. You pay with a $1 bill.

How much change do you get?

______

Show the change using Ⓠ, Ⓓ, and Ⓝ.

LESSON 9•7

Math Boxes

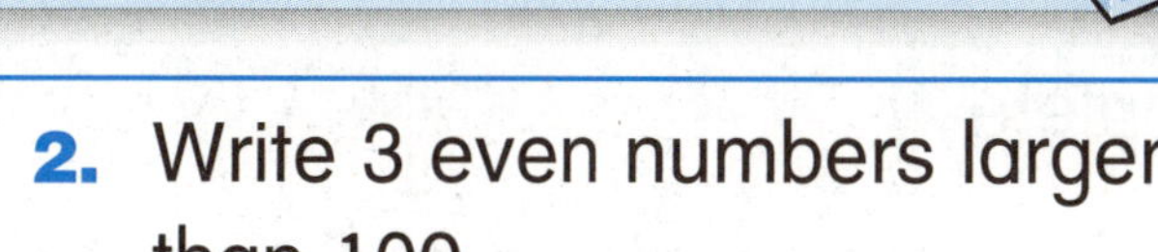

1. Draw two ways to show $\frac{2}{3}$.

2. Write 3 even numbers larger than 100.

_____, _____, _____

Write 3 odd numbers smaller than 100.

_____, _____, _____

3. Get 36 counters. Share them equally among 4 children.

How many counters does each child get? _____ counters

How many are left over?

_____ counters

4. I have $2.00. Can I buy 4 bags of chips for $0.55 each?

_____

5. Solve.

Unit

386 − 40 = _____

_____ = 198 − 60

259 − 40 = _____

_____ = 243 − 20

6. In 43,692, the value of

4 is _____________.

3 is _____________.

6 is _____________.

9 is _____________.

2 is _____________.

Date Time

Equivalent Units of Capacity

Complete.

U.S. Customary Units of Capacity

_____ pint = 1 cup

1 pint = _____ cups

_____ pints = 1 quart

_____ quarts = 1 half-gallon

_____ half-gallons = 1 gallon

Metric Units of Capacity

1 liter = _____ milliliters

$\frac{1}{2}$ liter = _____ milliliters

1. How many quarts are in 1 gallon? _____ quarts

2. How many cups are in 1 quart? _____ cups

In a half-gallon? _____ cups In 1 gallon? _____ cups

3. How many pints are in a half-gallon? _____ pints

In a gallon? _____ pints

"What's My Rule?"

4.

Rule

1 qt = 2 pt

qt	pt
2	
3	
	10
8	

5.

Rule

1 gal = 8 pt

gal	pt
2	
3	
	40
	80

Date Time

LESSON 9•8

Math Boxes

1. Write 5 names for 130.

MRB 16

2. Fill in the missing numbers.

MRB 98 99

3. Solve. Show your work.

$$\begin{array}{r} 49 \\ +\ 23 \\ \hline \end{array}$$

4.

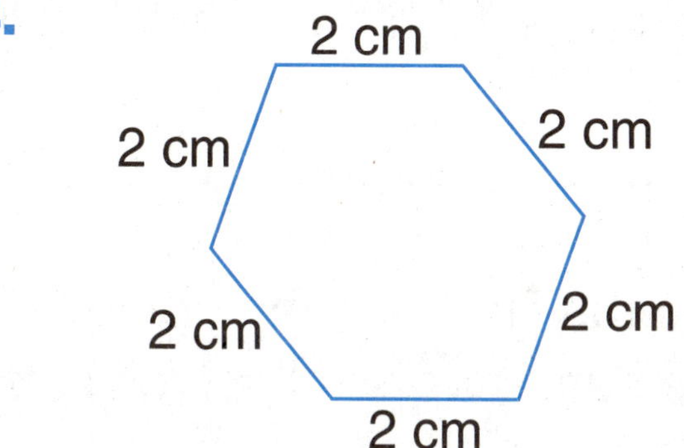

Perimeter = ______ cm

5. Estimate. Then solve.

Estimate:

______ + ______ = ______

$$\begin{array}{r} 57 \\ +\ 48 \\ \hline \end{array}$$

6. The total cost is $1.50. You pay with a $5 bill. How much change do you get?

Fill in the circle next to the best answer.

Ⓐ $6.50 Ⓑ $2.50

Ⓒ $4.00 Ⓓ $3.50

Weight

Weighing Pennies

Use a spring scale, letter scale, or diet scale to weigh pennies. Find the number of pennies that weigh about 1 ounce.

I found that ______ pennies weigh about 1 ounce.

Which Objects Weigh about the Same?

Work in a small group. Your group will be given several objects that weigh less than 1 pound.

1. Choose two objects. Hold one object in each hand and compare their weights. Try to find two objects that weigh about the same.

   Two objects that weigh about the same:

   ______________________ ______________________

2. After everyone in the group has chosen two objects that weigh about the same, weigh all the objects. Record the weights below.

Object	**Weight** (include unit)

Date Time

LESSON 9•9

Math Boxes

1. Use your Fraction Cards. Find another name for $\frac{3}{4}$. Circle the best answer.

A $\frac{1}{2}$ **B** $\frac{3}{8}$

C $\frac{6}{8}$ **D** $\frac{2}{3}$

2. Write 2 even 4-digit numbers.

________ ________

Write 2 odd 4-digit numbers.

________ ________

3. Use counters to solve.

18 orange slices are shared equally. Each child gets 4 slices.

How many children are sharing?

______ children

How many slices are left?

______ slices

4. Fill in the missing amount.

I had 38¢.

I spent ______¢.

I have 15¢ left.

5. Solve.

Unit

22 − 14 = ____

62 − 14 = ____

162 − 14 = ____

____ = 292 − 14

____ = 402 − 14

6. In 96,527, the value of

5 is ________.

6 is ________.

7 is ________.

2 is ________.

9 is ________.

Date Time

Math Boxes

1. I have 85¢. How many 5¢ pieces of candy can I buy?

_____ pieces of candy

2. 67,248

The value of 8 is __________.

The value of 2 is __________.

The value of 7 is __________.

The value of 6 is __________.

The value of 4 is ________.

3. How much money?

$10 $5
Ⓠ Ⓝ Ⓝ Ⓟ

$__________

4. The total cost of Neena's lunch is $7.50.

She paid with a $10 bill.

How much change will she get?

$ __________

5. Justin had $0.92 and found $0.21 more. How much does he have now? Estimate your answer and then solve.

Estimate:

________ + ________ = ________

Answer: __________

6. Use Ⓟ, Ⓝ, Ⓓ, and Ⓠ. Show $2.20.

Date Time

LESSON 10•1

Math Boxes

1. Double.

25¢ ________

55¢ ________

65¢ ________

85¢ ________

2. 15 children. $\frac{1}{3}$ are boys.

How many are boys? ________

How many are girls? ________

3. Count by 1000s.

________; 2,728; ________;

________; ________; ________;

________; ________; ________

4. Solve.

Unit

5 + 3 = ________

50 + 30 = ________

$$\begin{array}{r} 6 \\ +\ 3 \\ \hline \end{array} \qquad \begin{array}{r} 60 \\ +\ 30 \\ \hline \end{array}$$

5. Draw a rhombus. Make each side 2 cm long.

6. How many dots are in this 5-by-5 array?

________ dots in all

Good Buys Poster

Fruit/Vegetables Group

Seedless Grapes
99¢ lb

Carrots
1-lb bag
3/$1.00

Plums
69¢ lb

Oranges
$1.49 lb

Bananas
59¢ lb

Watermelons
$2.99 ea.

Celery
59¢ lb

Grain Group

Wheat Bread
16 oz
99¢

Saltines
1 lb
69¢

Hamburger Buns
16 oz
69¢

Meat Group

Pork & Beans
16 oz
2/89¢

Peanut Butter
18-oz jar
$1.29

Ground Beef
$1.99 lb

Chunk Light
Tuna
6.5 oz
69¢

Lunch Meat
1-lb package
$1.39

Milk Group

Gallon
Milk
$2.39

American
Cheese
8 oz
$1.49

6-pack
Yogurt
$2.09

Miscellaneous Items

Mayonnaise
32 oz
$1.99

Catsup
32 oz
$1.09

Grape Jelly
2-lb jar
$1.69

Date Time

Ways to Pay

Complete Problem 1. For Problems 2 and 3 choose two items from the Good Buys Poster on page 230. List the items and how much they cost in the table below.

For each item:

- Count out coins and bills to show several different ways of paying for each item.
- Record two ways by drawing coins and bills in the table. Use Ⓠ, Ⓓ, Ⓝ, Ⓟ, and $1.

Example:

You buy 1 pound of bananas. They cost 59¢ a pound. You pay with:
Ⓠ Ⓠ Ⓝ Ⓟ Ⓟ Ⓟ Ⓟ or Ⓓ Ⓓ Ⓓ Ⓓ Ⓓ Ⓝ Ⓟ Ⓟ Ⓟ Ⓟ

1. You buy *oranges*. Cost: $1.49 Pay with ____ or ____	**2.** You buy ________. Cost: ________ Pay with ____ or ____
3. You buy ________. Cost: ________ Pay with ____ or ____	**Try This** **4.** You buy ________ and ________. Cost: ________ Pay with ____ or ____

Date Time

LESSON 10•2

Word Values

Pretend the letters of the alphabet have the dollar values shown in the table. For example, the letter **g** is worth $7; the letter **v** is worth $22. The word **jet** is worth $10 + $5 + $20 = $35.

	a	b	c	d	e	f	g	h	i	j	k	l	m
Value	$1	$2	$3	$4	$5	$6	$7	$8	$9	$10	$11	$12	$13
	n	o	p	q	r	s	t	u	v	w	x	y	z
Value	$14	$15	$16	$17	$18	$19	$20	$21	$22	$23	$24	$25	$26

1. Which is worth more, **dog** or **cat**? ____________

2. Which is worth more, **whale** or **zebra?** ____________

3. How much is your first name worth? ____________

4. Write 2 spelling words you are trying to learn. Find their values.

   Word: ____________ Value: $____________

   Word: ____________ Value: $____________

5. What is the cheapest word you can make? It must have at least 2 letters.

   Word: ____________ Value: $____________

6. What is the most expensive word you can make?

   Word: ____________ Value: $____________

Try This

7. Think of the letter values as dimes. For example, **m** is worth 13 dimes; **b** is worth 2 dimes. Find out how much each word is worth.

   dog: $______ cat: $______ zebra: $______ whale: $______

   candy: $______ your last name: $____________

Date Time

LESSON 10•2

Math Boxes

1. Write $<$, $>$, or $=$.

1,257 ______ 2,157

7,925 ______ 5,297

10,129 ______ 1,129

2. Circle the answer.

\$2.88 is closer to:

\$2.80 or \$2.90

\$5.61 is closer to:

\$5.60 or \$5.70

\$1.97 is closer to:

\$1.90 or \$2.00

3. Put the heights in order. Find the median height.

Unit
inches

48 44 37 54 39

_____, _____, _____, _____, _____

The median height is

______ inches.

4. What is the temperature? Circle the best answer.

A. 85°F

B. 86°F

C. 83°F

D. 76°F

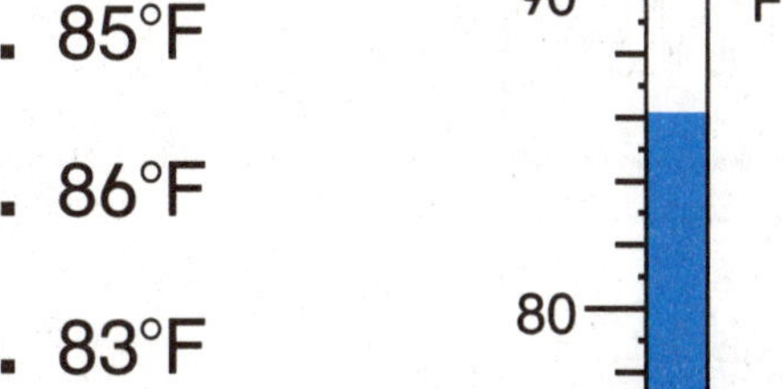

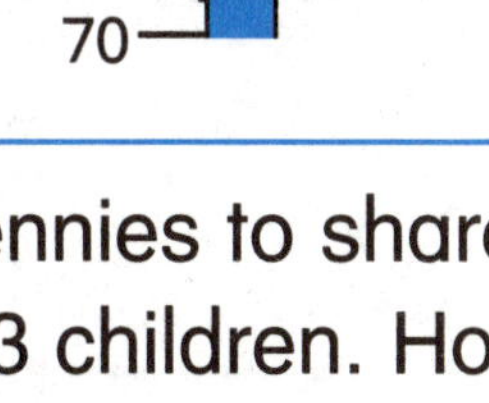

5. Draw the hour and minute hands to show the time 20 minutes later than 6:15.

What time does the clock show now?

_____:_____

6. You have 21 pennies to share equally among 3 children. How many pennies does each child get?

______ pennies

How many are left over?

______ pennies

Date Time

Calculator Dollars and Cents

To enter $4.27 into your calculator, press (4) (.) (2) (7).

To enter 35¢ into your calculator, press (.) (3) (5).

1. Enter $3.58 into your calculator. The display shows ____________.

2. Enter the following amounts into your calculator.

   Record what the display shows.
   Don't forget to clear between each entry.

Price	Display
$2.75	____________
$1.69	____________
$12.32	____________

Make up prices that are more than $1.00.

3. Enter 68¢ into your calculator. The display shows ____________.

Calculator Dollars and Cents *continued*

4. Enter the following amounts into your calculator. Record what you see in the display.

Make up prices that are less than $1.00.

5. Use your calculator to add $1.55 and $0.25.

What does the display show? ______

Explain what happened. ______________________

____

____

Date Time

LESSON 10•3

Pick-a-Coin Directions

Materials ☐ 1 die ☐ calculator for each player

☐ *Pick-a-Coin* record table for each player (*Math Journal 2,* p. 237 or *Math Masters*, p. 469)

Players 2 to 4

Skill Add dollar bill and coin combinations

Object of the Game To add the largest value

Summary

Players roll a die. The numbers that come up are used as numbers of coins and dollar bills. Players try to make collections of coins and bills with the largest value.

Directions

Take turns. When it is your turn, roll the die five times. After each roll, record the number that comes up on the die in any one of the empty cells in the row for that turn on your record table. Then use a calculator to find the total amount for that turn. Record the total in the table.

After four turns, use your calculator to add the four totals. The player with the largest Grand Total wins.

Example: On his first turn, Brian rolled 4, 2, 4, 1, and 6. He filled in his record table like this:

***Pick-a-Coin* Record Table**

	Ⓟ	Ⓝ	Ⓓ	Ⓠ	$1	Total
1st turn	2	1	4	4	6	$7.47
2nd turn						$___.____
3rd turn						$___.____
4th turn						$___.____
				Grand Total		$___.____

Date Time

Pick-a-Coin Record Tables

	Ⓟ	Ⓝ	Ⓓ	Ⓠ	$1	Total
1st turn						$___ . ______
2nd turn						$___ . ______
3rd turn						$___ . ______
4th turn						$___ . ______
					Grand Total	$___ . ______

	Ⓟ	Ⓝ	Ⓓ	Ⓠ	$1	Total
1st turn						$___ . ______
2nd turn						$___ . ______
3rd turn						$___ . ______
4th turn						$___ . ______
					Grand Total	$___ . ______

	Ⓟ	Ⓝ	Ⓓ	Ⓠ	$1	Total
1st turn						$___ . ______
2nd turn						$___ . ______
3rd turn						$___ . ______
4th turn						$___ . ______
					Grand Total	$___ . ______

Date Time

LESSON 10•3 Finding the Median

One way to find the median:

◆ Circle the tally marks.

Example:

Arm Span (inches)	Frequency	
	Tallies	Number
42	ⓘⓘ	2
43		
44	ⓘ	1
45		
46	ⓘ/ /ⓘ	4
47	ⓘⓘ	2
48		
49	ⓘ	1
Total		

Median 46 inches

Another way:

◆ Order the arm-span data in inches.

Example:

~~42~~, ~~42~~, ~~44~~, ~~46~~, 46, 46, ~~46~~, ~~47~~, ~~47~~, ~~49~~

Median 46 inches

Find the median in two ways. Show your work.

1. One way:

Number of Siblings	Frequency	
	Tallies	Number
6		
5	/	1
4		
3	/ / / /	4
2	/ /	2
1	/ / / /	4
0	/	1
Total		12

Median ______ siblings

2. Another way:

0 1 1 1 1 2 2 3 3 3 3 5

Median ______ siblings

Date Time

LESSON 10•3

Math Boxes

1. ______ pennies = \$3.00

______ nickels = \$3.00

______ dimes = \$3.00

______ quarters = \$3.00

2. Count 20 pennies.

$\frac{1}{2}$ = ______ pennies

$\frac{1}{4}$ = ______ pennies

$\frac{1}{5}$ = ______ pennies

3. Complete the frames.

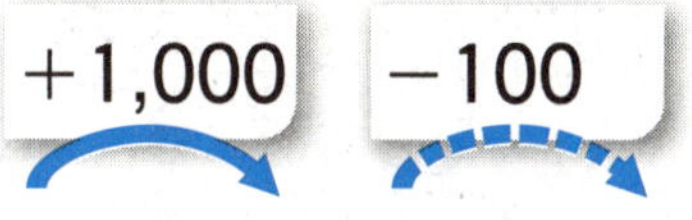

2,463		
		4,363

4. Solve.

Unit

9 − 5 = ______

______ = 90 − 50

900 − 500 = ______

______ = 9,000 − 5,000

5. Match.

5 ft	3 yd
24 in.	60 in.
9 ft	2 ft

6. Draw an 8-by-4 array.

How many in all? ______

Date Time

LESSON 10•4 Then-and-Now Poster

1897

Cheddar Cheese
$\frac{1}{2}$ lb
6¢

Crackers
1 lb
6¢

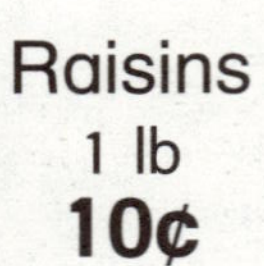

Raisins
1 lb
10¢

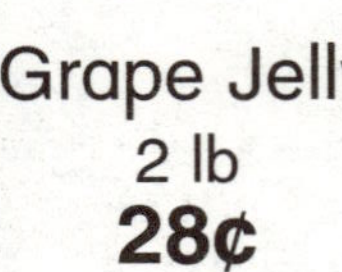

Grape Jelly
2 lb
28¢

Catsup
32 oz/1 qt
25¢

20-Inch Girl's Bicycle
$29.00

Child's Wagon
Large Size –15" × 30"
$1.65

Harmonica
Ten Double Holes
45¢

Now

Cheddar Cheese
8 oz ($\frac{1}{2}$ lb)
$2.99

Crackers
1 lb
$2.49

Raisins
1 lb
$2.39

Grape Jelly
2 lb
$2.29

Catsup
32 oz / 1 qt
$2.79

20-Inch Girl's Bicycle
$119.99

Child's Wagon
Medium Size – $15\frac{1}{2}$" × 34"
$47.99

Harmonica
Ten Double Holes
$17.50

Then-and-Now Prices

Use your calculator.

1. How much did a 20-inch bicycle cost in 1897? ____________

   How much does it cost now? ____________

   How much more does it cost now? ____________

2. How much more does a pound of cheese cost now

   than it did in 1897? ____________

3. In 1897, raisins were packed in cartons. Each carton contained 24 one-pound boxes. How much did a

   24-pound carton cost then? ____________

   How much would it cost now? ____________

4. Which item had the biggest price increase from then to now?

   ____________________________ had the biggest price increase.

   How much more does it cost now? ____________

5. Our Own Problems about Then-and-Now:

   ____

   ____

   ____

   ____

   ____

   ____

Date Time

LESSON 10•4 Math Boxes

1. Write <, >, or =.

1,292 + 10 ____ 1,285 + 15

3,791 + 7 ____ 3,799 + 7

5,020 + 100 ____ 5,125 + 25

2. Fill in the blanks to estimate the total cost.

$2.43 + $0.39 is about

______ + ______ = ______

$0.88 + $0.67 is about

______ + ______ = ______

3. Arrange the numbers in order.

Find the median.

98 56 143 172 81

______, ______, ______,

______, ______

The median is ______.

4. Show 55°F.

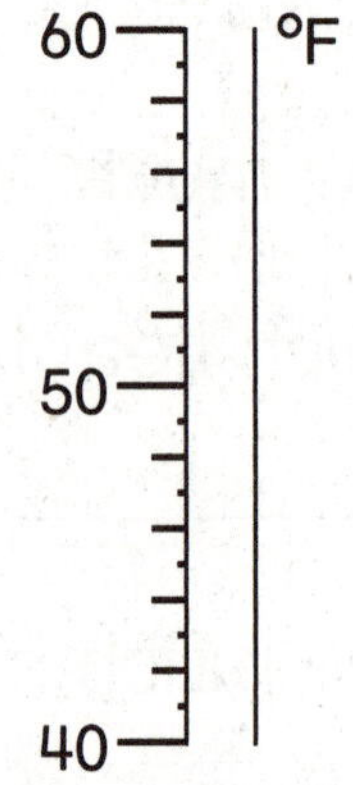

5. It is 6:15. Draw the hour and minute hands to show the time 15 minutes later.

What time does the clock show?

___:___

6. Use counters to solve.

35 blocks are shared equally among 3 children. How many blocks does each child get?

______ blocks

How many blocks are left over?

______ blocks

Date Time

Estimating and Buying Food

Choose items to buy from the Good Buys Poster on journal page 230. For each purchase:

- Record the items on the sales slip.
- Write the price of each item on the sales slip.
- Estimate the total cost and record it.
- Your partner then uses a calculator to find the exact total cost and writes it on the sales slip.

Purchase 1	**Price**
Items:	
______________	\$ ___.___
______________	\$ ___.___
Estimated cost: about	\$ 1.60
Exact total cost:	\$ 1.58

Purchase 2	**Price**
Items:	
______________	\$ ___.___
______________	\$ ___.___
Estimated cost: about	\$ ___.___
Exact total cost:	\$ ___.___

Purchase 3	**Price**
Items:	
______________	\$ ___.___
______________	\$ ___.___
Estimated cost: about	\$ ___.___
Exact total cost:	\$ ___.___

LESSON 10•5

Math Boxes

1. Show $1.73 in two different ways. Use Ⓟ, Ⓝ, Ⓓ, and Ⓠ.

____

____

____

____

2. Draw a picture of 10 children.

$\frac{1}{2}$ play ball. How many? ____

$\frac{3}{10}$ jump rope. How many? ____

$\frac{1}{5}$ skate. How many? ____

3. Fill in the rule and the missing numbers.

Rule

in	out
1,342	2,342
3,019	4,019
4,650	
	6,700

4. Solve.

Unit
km

6 + 5 = ________

60 + 50 = ________

600 + 500 = ________

6,000 + 5,000 = ________

5. Write <, >, or =.

1 qt ____ 1 pt

3 c ____ 1 gal

1 qt ____ 4 c

1 gal ____ 5 pt

6. There are 3 drink boxes per pack. How many packs are needed to serve 25 second graders and 2 teachers one drink box each? Draw an array. Circle the best answer.

_____ packs are needed.

A. 30 **B.** 8 **C.** 9 **D.** 10

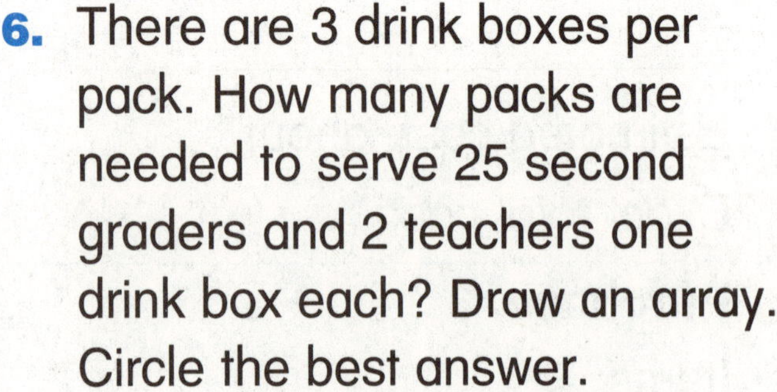

Date _______________ Time _______________

LESSON 10•6 Math Boxes

1. What number is shown by the blocks?

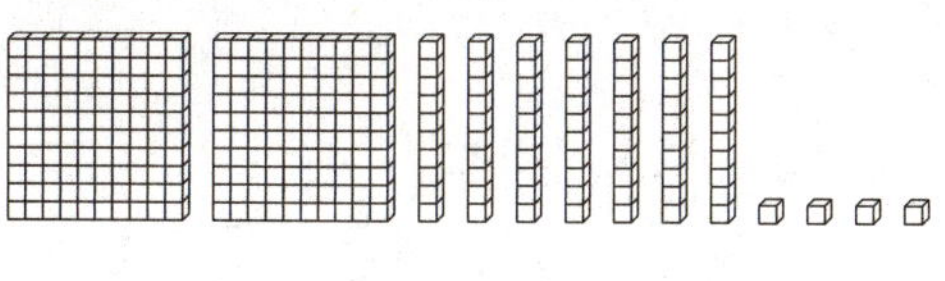

____

2. Joe has $1.00 and spends 65¢. How much change will he get?

____

3. What is the temperature?

_______ °F

Is it warm or cold outside?

____

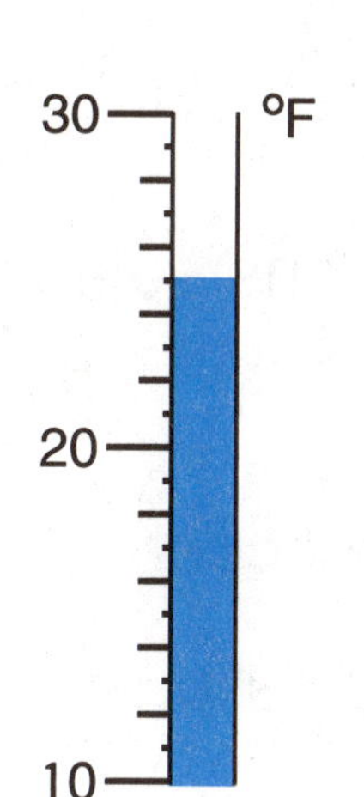

4. You buy some stickers for $1.89. Show 2 ways to pay. Use Ⓟ, Ⓝ, Ⓓ, Ⓠ, and [$1].

5. Write the names of 3 objects that are shaped like cylinders.

____

____

____

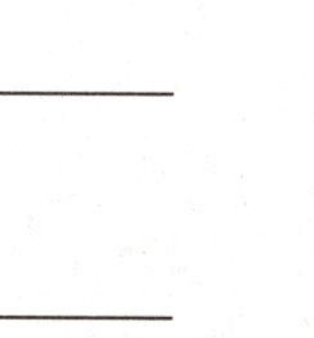

6. How many stars in all?

_______ stars

Fill in the multiplication diagram.

rows	stars per row	stars in all

Date Time

Making Change

Work with a partner. Use your tool-kit coins and bills. One of you is the shopper. The other is the clerk.

The shopper does the following:

- Chooses one item from each food group on the Good Buys Poster on journal page 230.
- Lists these items on the Good Buys sales slip in the shopper's journal on page 247.
- Writes the cost of each item on the sales slip.
- Estimates the total cost of all the items and writes it on the sales slip.
- Pays with a $10 bill.
- Estimates the change and writes it on the sales slip.

The clerk does the following:

- Uses a calculator to find the exact total cost.
- Writes the exact total cost on the sales slip.
- Gives the shopper change by counting up.
- Writes the exact change from $10.00 on the sales slip.

Change roles and repeat.

Making Change *continued*

The Good Buys Store Sales Slip

	Item	Cost
Fruit/vegetables group	______	$ ___ . ____
Grain group	______	$ ___ . ____
Meat group	______	$ ___ . ____
Milk group	______	$ ___ . ____
Miscellaneous items	______	$ ___ . ____

Estimated total cost $ ___ . ____

Estimated change from $10.00 $ ___ . ____

Exact total cost $ ___ . ____

Exact change from $10.00 $ ___ . ____

Date Time

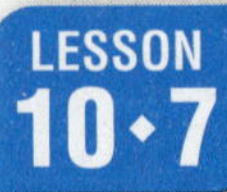

LESSON 10•7

The Area of My Handprint

The area of each □ is 1 square centimeter. Other ways to write *square centimeter* are sq cm and cm^2.

The area of my handprint is _____ square centimeters, or _____ sq cm.

Date Time

The Area of My Footprint

The area of each ☐ is 1 square centimeter. Other ways to write *square centimeter* are sq cm and cm^2.

The area of my footprint is _____ square centimeters, or _____ sq cm.

Worktables

Use a Pattern-Block Template to draw each of the different-size and different-shape worktables you made. Write the name of each shape next to your drawing.

Date Time

Geoboard Dot Paper

1.

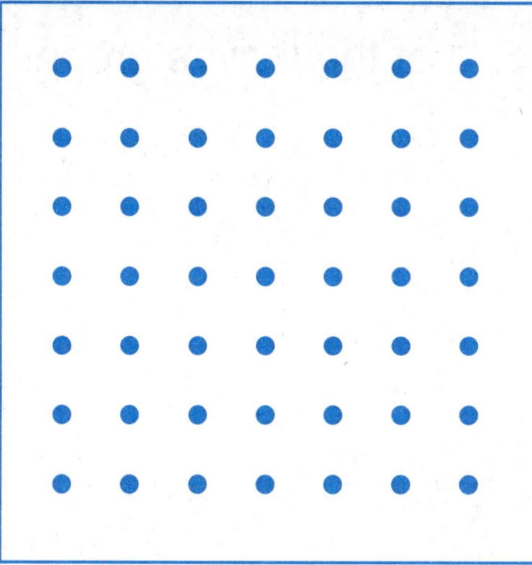

2.

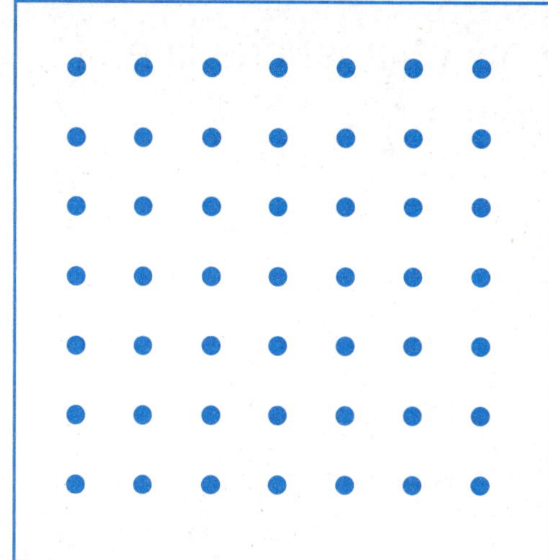

3.

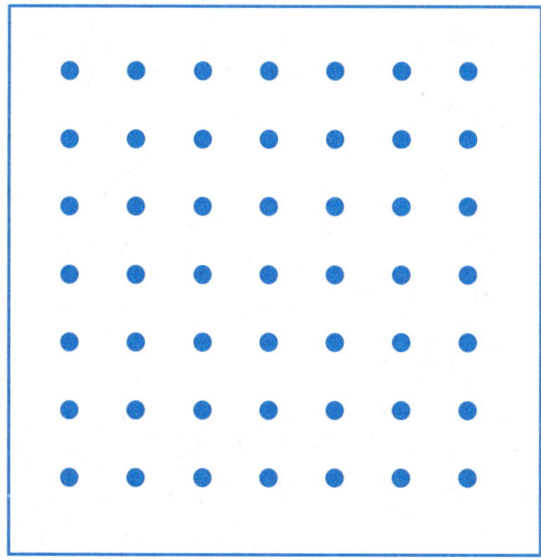

4.

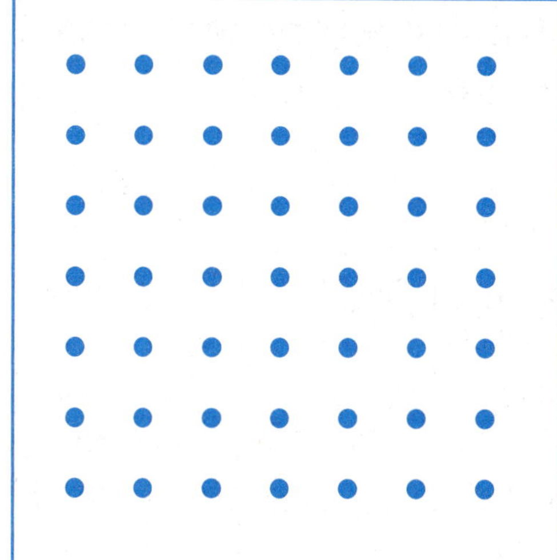

5.

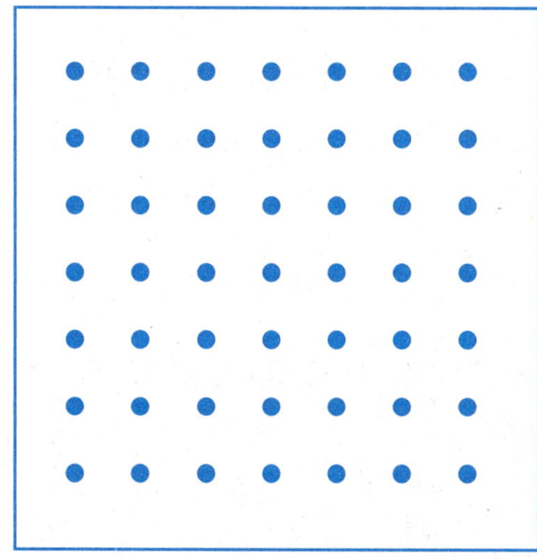

6.

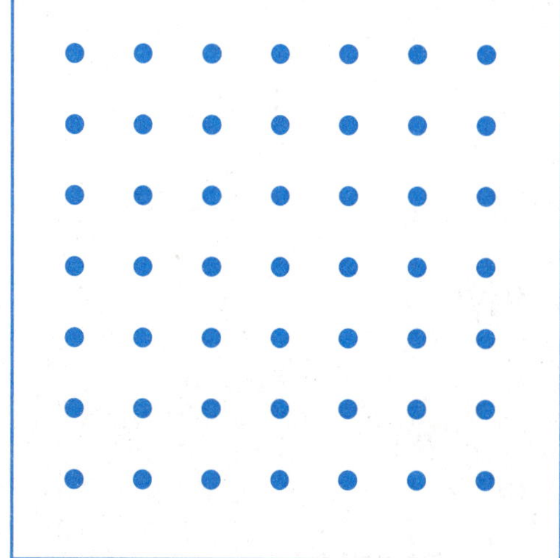

LESSON 10·7

Math Boxes

1. Draw at least one line of symmetry.

2. Color $\frac{2}{3}$ of the leaves green.

3. Write as dollars and cents. Eight dollars and forty-three cents:

____

fifteen dollars and 6 cents:

____

fifty dollars and seventeen cents: _______

4.

in	out
32	
56	
45	
	97
89	

Rule

+9

5. Write <, >, or =.

1 hour ____ 30 minutes

3 months ____ 1 year

7 days ____ 1 week

6. Use counters to solve.
15 marbles are shared equally. Each child gets 6 marbles. How many children are sharing?

_______ children

How many marbles are left over?

_______ marbles

Date Time

Money Exchange Game Directions

Materials
- ☐ 1 six-sided die
- ☐ 1 ten- or twelve-sided die
- ☐ 24 pennies, 39 dimes, thirty-nine $1 bills, and one $10 bill per player

Players 2 or 3

Skill Make exchanges between coins and bills

Object of the Game To be the first to trade for $10

Directions

1. Each player puts 12 pennies, 12 dimes, twelve $1 bills, and one $10 bill in the bank.
2. Players take turns. Players use a six-sided die to represent pennies. Players use a ten- or twelve-sided die to represent dimes.
3. Each player
   - rolls the dice.
   - takes from the bank the number of pennies and dimes shown on the faces of the dice.
   - puts the coins in the correct columns on his or her Place-Value Mat on journal page 254.
4. Whenever possible, a player replaces 10 coins or bills of a lower denomination with a coin or bill of the next higher denomination.
5. The first player to trade for a $10 bill wins.

If there is a time limit, the winner is the player with the largest number on the mat when time is up.

Date Time

LESSON 10•8

Place-Value Mat

$10	$1 dollars	(D) dimes	(P) pennies
1,000s	100s	10s	1s

Ballpark Estimates

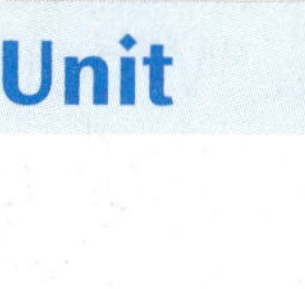

Fill in the unit box. Then, for each problem:

Make a ballpark estimate before you add.

Write a number model for your estimate.

Use your calculator and solve the problem. Write the exact answer in the box.

Compare your estimate to your answer.

1. Ballpark estimate: ______ 148 + 27	**2.** Ballpark estimate: ______ 163 + 32	**3.** Ballpark estimate: ______ 133 + 35
4. Ballpark estimate: ______ 143 + 41	**5.** Ballpark estimate: ______ 184 + 23	**6.** Ballpark estimate: ______ 154 + 83

Date Time

LESSON 10•8

Math Boxes

1. What number is shown by the blocks?

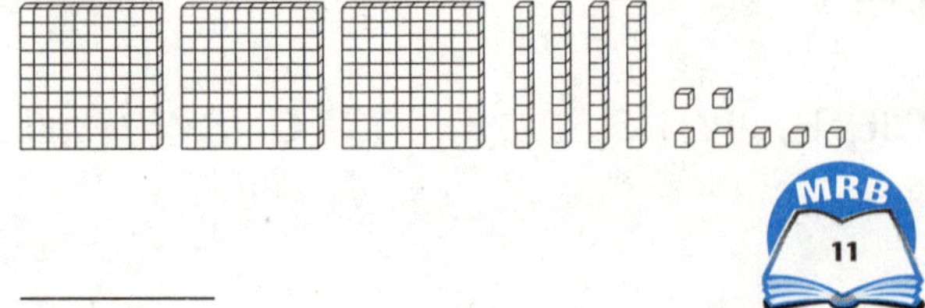

______

MRB 11

2. Kelly has $10. Her lunch total was $8.75. How much change will she get?

______

3. In the morning, it was 62°F. By afternoon, the temperature was 75°F. How much did the temperature rise? ______

Start | Change → | End

Number model: ______________

4. Cross out the names that don't belong.

10¢

ten cents, $\frac{1}{10}$ of a dollar,

$10.00, Ⓓ, ⓃⓃ, $0.01,

$\frac{1}{100}$ of a dollar,

$\frac{1}{2}$ of a dollar

5. Which object is shaped like a cone? Circle the best answer.

A shoe box

B party hat

C paper towel roll

D globe

6. 4 ladybugs. 5 spots on each ladybug. How many spots?

______

Fill in the diagram and write a number model.

lady bugs	spots per lady bug	spots in all

MRB 112 113

______ × ______ = ______

Date Time

LESSON 10·9

Math Boxes

1. Use your Pattern-Block Template.

   Trace a trapezoid. Draw the line of symmetry.

2. Color $\frac{3}{4}$ of the circle.

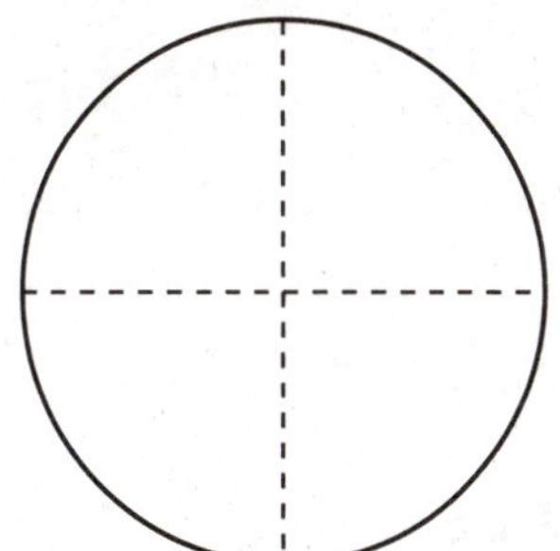

What fraction of the circle is not colored? ______

3. Write the amounts.

   Five thousand six hundred eight dollars and twelve cents

   ____

   Two hundred sixteen dollars and sixty-eight cents

   ____

   Three hundred nine dollars and five cents

   ____

4. **Rule**

   12 in. = 1 ft

in	out
6	
	2
48	

5. ______ hours in a day

   ______ days in a week

   ______ months in a year

   ______ weeks in a year

6. Use counters to solve.
   26 children. 2 children for each computer. How many computers?

   ______ computers

Date Time

Place Value

1. Match names.

A. 5 ones ______ 50

B. 5 tens ______ 500

C. 5 hundreds ______ 50,000

D. 5 thousands ______ 5

E. 5 ten-thousands ______ 5,000

Fill in the blanks. Write ones, tens, hundreds, thousands, or ten-thousands.

2. The 7 in 187 stands for 7 ______________.

3. The 2 in 2,785 stands for 2 ______________.

4. The 3 in 4,239 stands for 3 ______________.

5. The 0 in 13,409 stands for 0 ______________.

6. The 5 in 58,047 stands for 5 ______________.

Continue.

7. 364; 365; 366; ________; ________; ________

8. 996; 997; 998; ________; ________; ________

9. 1,796; 1,797; 1,798; ________; ________; ________

10. 1,996; 1,997; 1,998; ________; ________; ________

11. 9,996; 9,997; 9,998; ________; ________; ________

Date Time

LESSON 10·10

Math Boxes

1. Circle the digit in the 1,000s place.

4, 6 9 4

2 9, 4 0 0

2 0, 0 0 4

5, 0 1 9

Read each number to a partner.

2. I have a 5-dollar bill. I spend \$4.38. How much change do I get?

____

3. Show 25°C on the thermometer.

Is it good weather to go ice skating or to go to the beach?

____

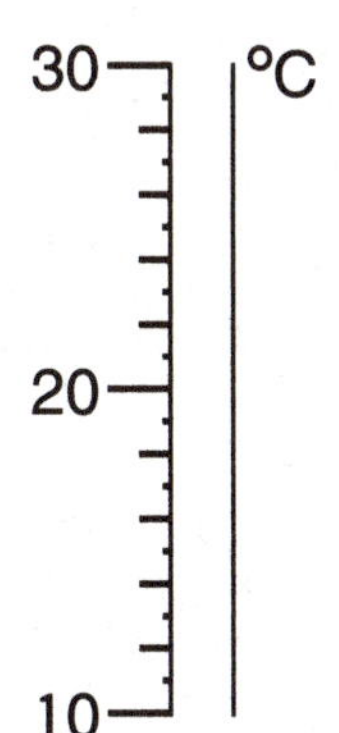

4. Write 5 names for \$0.75.

____

____

____

____

____

5. Write the names of 3 objects shaped like rectangular prisms.

____

____

____

6. 5 wagons. 4 wheels on each wagon. How many wheels?

______ wheels

wagons	wheels per wagon	wheels in all

MRB 112 113

Date Time

LESSON 10•11

Parentheses Puzzles

Parentheses can make a big difference in a problem.

Example:

$15 - 5 + 3 = ?$

$(15 - 5) + 3 = (10) + 3 = 13$; but

$15 - (5 + 3) = 15 - (8) = 7$

Solve problems containing parentheses.

1. $7 + (8 - 3) =$ _____
2. $(4 + 11) - 6 =$ _____
3. $8 + (13 - 9) =$ _____
4. _____ $= (12 + 8) - 16$
5. $140 - (20 + 80) =$ _____
6. _____ $= (30 + 40) - 70$

Put in parentheses to solve the puzzles.

7. $12 - 4 + 6 = 14$
8. $15 - 9 - 4 = 10$
9. $140 - 60 + 30 = 110$
10. $500 = 400 - 100 + 200$
11. $3 \times 2 + 5 = 11$
12. $2 \times 5 - 5 = 0$

Date Time

LESSON 10•11

Math Boxes

1. Choose any shape on your Pattern-Block Template that has at least 2 lines of symmetry. Trace the shape and draw the lines of symmetry.

2. Circle $\frac{3}{12}$.

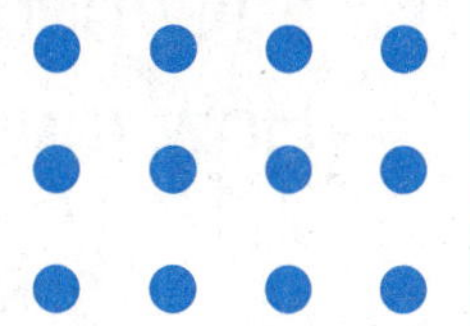

What fraction of dots is not circled? Circle the best answer.

A $\frac{9}{3}$ **B** $\frac{9}{12}$

C $\frac{6}{12}$ **D** $\frac{3}{9}$

Write another name for this fraction ________.

3. Use your calculator to find the total.

[$1] [$1] [$1] [$1] = $________

Ⓠ Ⓠ Ⓠ = $________

Ⓓ Ⓓ Ⓓ Ⓓ Ⓓ = $________

Ⓝ Ⓝ Ⓝ Ⓝ Ⓝ Ⓝ Ⓝ = $________

Total = $________

4.

Rule
5Ⓝ = 1Ⓠ

Ⓝ	Ⓠ
	3
	4
30	
	10
	20

5. ________ months = 1 year

________ months = 2 years

________ months = 3 years

________ months = 4 years

6. You have 18 pieces of gum to share equally. If each child gets 4 pieces, how many children are sharing?

________ children

How many pieces of gum are left over?

________ pieces of gum

Math Boxes

1. 6 lily pads. 3 frogs per lily pad. How many frogs in all?

_____ frogs

Fill in the diagram and write a number model.

lily pads	frogs per lily pad	frogs in all

_____ × _____ = _____

2. Draw a 7-by-3 array.

How many in all? _____

3. 17 magazines are shared equally among 5 children. Draw a picture to help you.

Each child gets ___ magazines.

There are ___ magazines left.

4. 8 books per shelf. 4 shelves. Fill in the diagram and solve.

shelves	books per shelf	books in all

There are _____ books.

5. This is a _____ -by- _____ array.

How many dots in all?

_____ dots

6. 11 balloons are shared equally among 4 children. How many balloons does each child get?

_____ balloons

How many are left over?

_____ balloons

Date Time

Math Boxes

1. Circle the unit that makes sense.

   Grandma's house is

   5 ______ away. km dm

   Mona's goldfish is

   8 ______ long. cm m

   Ahmed's dad is

   68 ______ tall. cm in.

2. Circle the fraction that is bigger. Use your Fraction Cards to help.

   $\frac{2}{3}$ or $\frac{2}{2}$ $\frac{4}{5}$ or $\frac{2}{5}$

   $\frac{2}{8}$ or $\frac{5}{6}$ $\frac{3}{6}$ or $\frac{1}{4}$

3. Write a number with 5 in the thousands place.

   ____

   What is the value of the digit 5 in your number?

   ______

4. Draw a 3-by-6 array.

   How many in all? ______

5. Had a $10 bill.
   Spent $8.90.
   How much change?

   ______

6. Write the fact family.

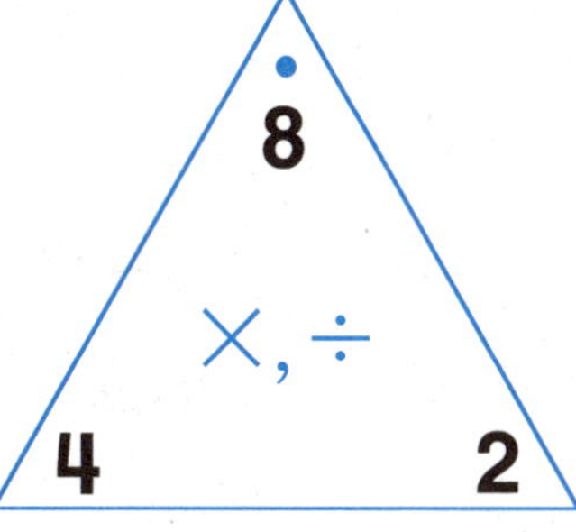

   ______ × ______ = ______

   ______ × ______ = ______

   ______ ÷ ______ = ______

   ______ ÷ ______ = ______

MRB 38

Art Supply Poster

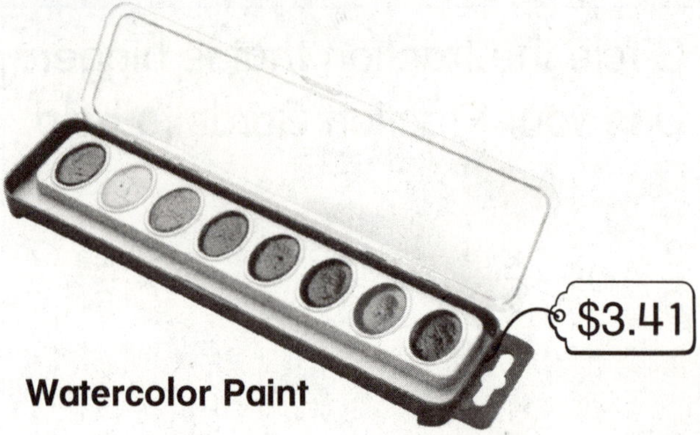

$3.41
Watercolor Paint

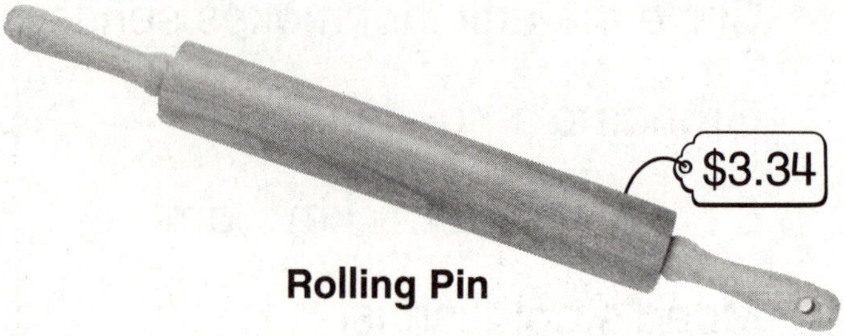

$3.34
Rolling Pin

$0.84
Glue

CLAY
$4.26
Modeling Clay

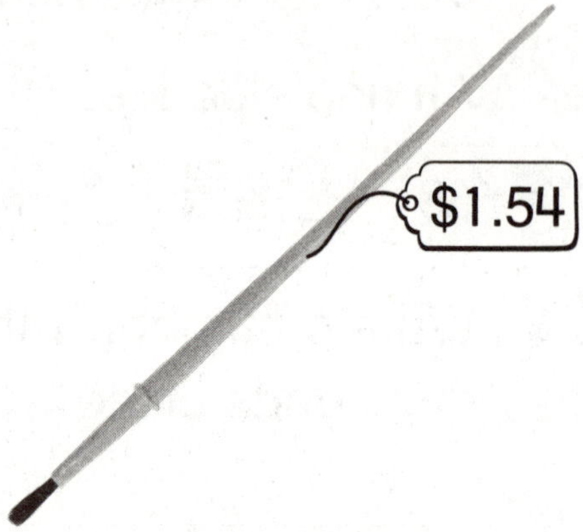

$1.54
Paintbrush

CONSTRUCTION PAPER
$0.65
Construction Paper

PIPE CLEANERS 
$0.76
Pipe Cleaners

$2.23
Scissors

8 COLOR MARKERS
$1.37
Color Markers

Date Time

Buying Art Supplies

Estimate the total cost for each pair of items.

Write your estimate in the answer space.

Add to find the total cost.

Check your estimate with your total cost.

1. pipe cleaners and watercolors Estimated Cost: ____ Total Cost: ____	**2.** clay and construction paper Estimated Cost: ____ Total Cost: ____	**3.** paintbrush and scissors Estimated Cost: ____ Total Cost: ____
4. glue and construction paper Estimated Cost: ____ Total Cost: ____	**5.** markers and glue Estimated Cost: ____ Total Cost: ____	**6.** clay and rolling pin Estimated Cost: ____ Total Cost: ____

LESSON 11•2

Comparing Costs

Use the Art Supply Poster on journal page 264.
In Problems 1–6, circle the item that costs more.
Then find how much more.

1. glue or markers How much more? _______	**2.** construction paper or paintbrush How much more? _______
3. pipe cleaners or paintbrush How much more? _______	**4.** rolling pin or scissors How much more? _______
5. watercolors or markers How much more? _______	**6.** paintbrush or watercolors How much more? _______

7. You buy a pack of construction paper. You pay with a $1 bill.

Should you get more or less than 2 quarters in change? _______

8. You buy pipe cleaners. You pay with a $1 bill.

How much change should you get? _______

9. You buy a rolling pin. You pay with a $5 bill.

How much change should you get? _______

Date Time

Data Analysis

Lily collected information about the ages of people in her family.

Complete.

Name	Age
Dave	40
Dedra	36
Jamal	12
Tyler	10
Lily	8

1. The oldest person is ______________.

   Age: ________ years

2. The youngest person is ______________.

   Age: ________ years

3. Range of the ages (oldest minus the youngest):

   ________ years

4. Middle value of the ages: ________ years

Fill in the blanks with the name of the correct person.

5. Jamal is about 4 years older than ______________.

6. Tyler is about 26 years younger than ______________.

7. Dedra is about 3 times as old as ______________.

8. Dave is about 4 times as old as ______________.

Try This

9. Tyler is about $\frac{1}{4}$ as old as ______________.

Date Time

LESSON 11·2

Math Boxes

1. Solve and show your work.

$$\begin{array}{r} 47 \\ -\ 39 \\ \hline \end{array}$$

Unit
students

2. What is the minimum number (the smallest) in the list?

____

2,371; 429; 578; 1,261

3. What shape is a globe? Circle the best answer.

A. sphere

B. rectangular prism

C. cone

D. cylinder

4. 20 campers divided equally among 5 tents. How many campers in each tent?

_______ campers

5. Fill in the missing numbers.

		1,065	
	1,074		

6. What is the range of this set of numbers (the largest minus the smallest)?

____

75, 93, 108, 52

Date Time

LESSON 11·3

Trade-First Subtraction

- Make a ballpark estimate for each problem and write a number model for your ballpark estimate.
- Use the trade-first method of subtraction to solve each problem.

Example:

Ballpark estimate:

40 − 20 = 20

longs 10s	cubes 1s
2	17
~~3~~	~~7~~
− 1	9
1	8

Answer 18

1. Ballpark estimate:

____

longs 10s	cubes 1s
2	8
− 1	9

Answer

2. Ballpark estimate:

____

longs 10s	cubes 1s
3	1
− 1	7

Answer

3. Ballpark estimate:

____

longs 10s	cubes 1s
7	6
− 5	9

Answer

4. Ballpark estimate:

____

longs 10s	cubes 1s
3	5
− 2	6

Answer

5. Ballpark estimate:

____

longs 10s	cubes 1s
4	4
− 2	7

Answer

LESSON 11·3

Math Boxes

1. The perimeter is about

_____ cm.

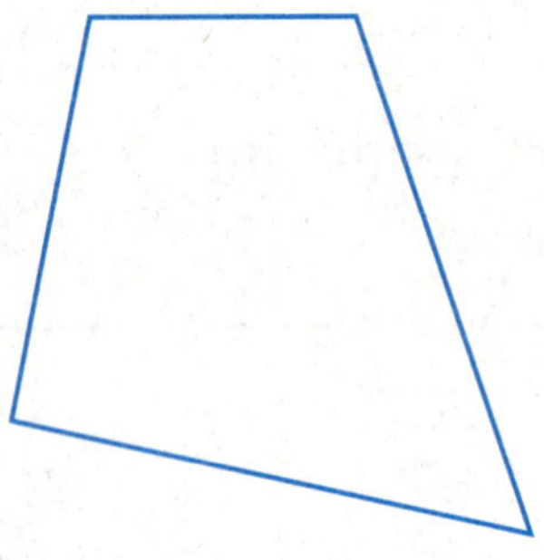

2. Divide into:

halves fourths

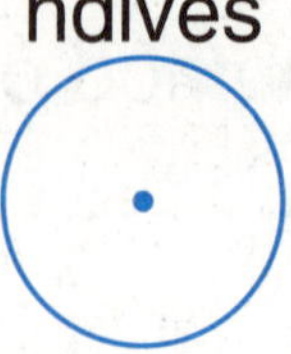

Write <, >, or =.

$\frac{1}{2}$ _____ $\frac{1}{4}$ $\frac{2}{4}$ _____ $\frac{1}{2}$

$\frac{1}{2}$ _____ $\frac{3}{4}$

3. What is the value of the digit 4 in each number?

14 ________

142 ________

436 ________

4,678 ________

4.

________-by-________ array

How many in all? ________

5. I had a 10-dollar bill. I spent $5.23. How much change did I receive? Fill in the circle next to the best answer.

Ⓐ $3.80 Ⓑ $4.77

Ⓒ $5.00 Ⓓ $15.23

6. Complete the Fact Triangle. Write the fact family.

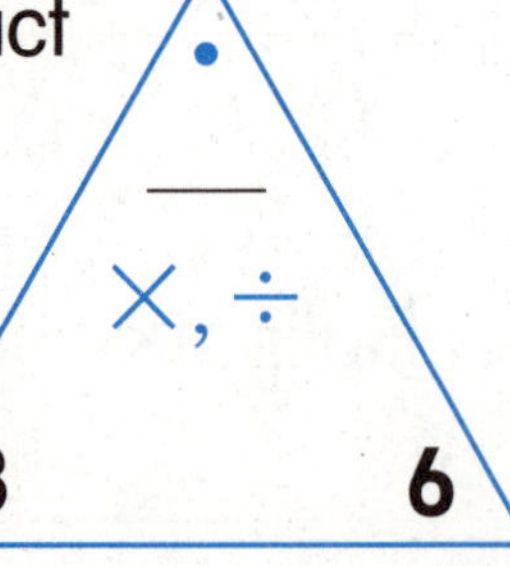

_____ × _____ = _____

_____ × _____ = _____

_____ ÷ _____ = _____

_____ ÷ _____ = _____

LESSON 11•4

Math Boxes

1. Solve and show your work.

$$\begin{array}{r} 71 \\ -\ 23 \\ \hline \end{array}$$

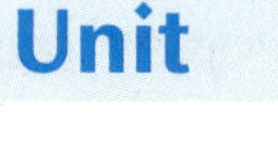

2. What is the maximum number (the largest) in the list?

______

7,946; 2,599; 17,949; 8,112

3. What shape is a can of soup?

______

4. 15 baseball cards are shared equally among 4 children. How many cards does each child get?

____ cards

How many left over? ______

5. Fill in the missing numbers.

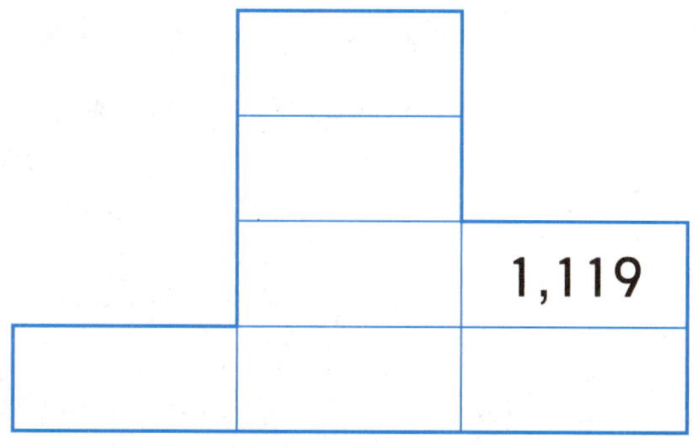

6. What is the range of this list of numbers (the largest minus the smallest)?

______

29, 132, 56, 30

Date Time

Multiplication Number Stories

1. 4 insects on the flower. How many legs in all?

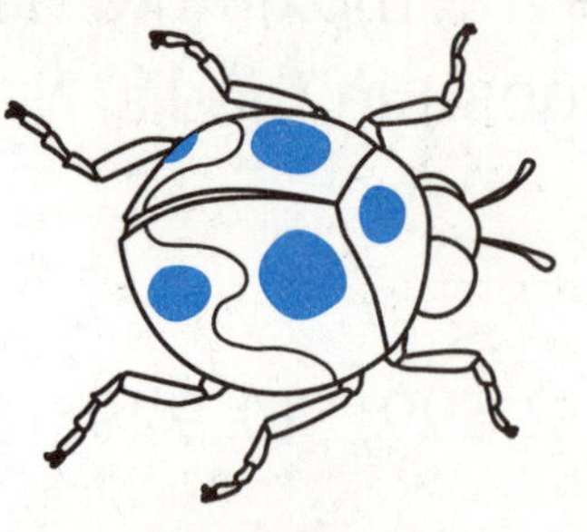

Has 6 legs

insects	legs per insect	legs in all

Answer: _____ legs

Number model: _____ × _____ = _____

2. 3 vans full of people. How many people in all?

Holds 10 people

vans	people per van	people in all

Answer: _____ people

Number model: _____ × _____ = _____

3. 9 windows. How many panes in all?

Has 4 panes

windows	panes per window	panes in all

Answer: _____ panes

Number model: _____ × _____ = _____

Date Time

Multiplication Number Stories *continued*

Try This

Use the pictures to make up two multiplication number stories.

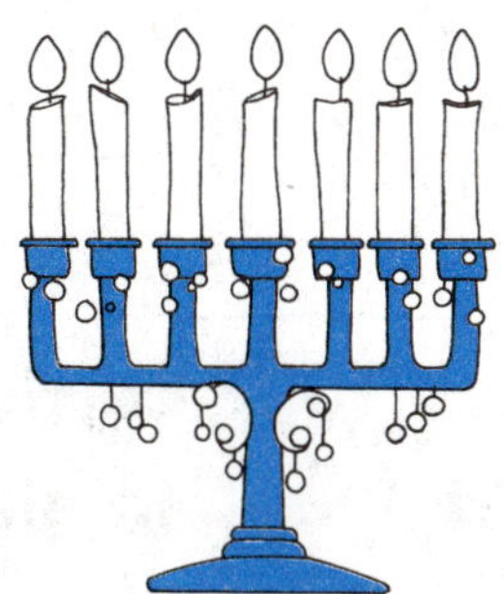

Has 7 candles

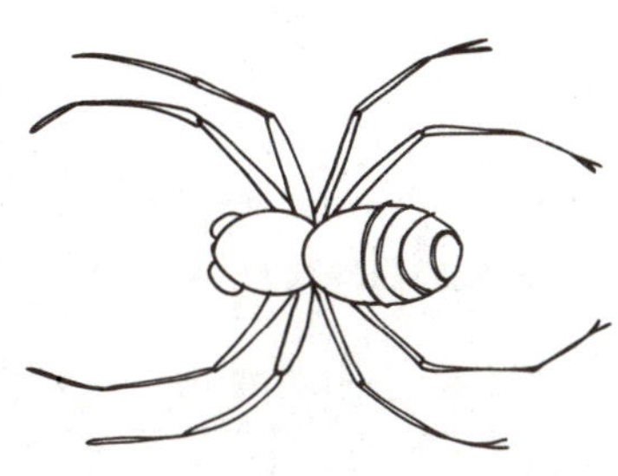

Has 8 legs

Has 5 players

For each story:

- Fill in the multiplication diagram.
- Draw a picture or array and find the answer.
- Fill in the number model.

4. ______________________

____

____

________	________ per ________	________ in all

Answer: _____

Number model: _____ × _____ = _____

5. ______________________

____

____

________	________ per ________	________ in all

Answer: _____

Number model: _____ × _____ = _____

LESSON 11•5 Division Number Stories

For each number story:

- Fill in the diagram.
- On a separate sheet of paper, draw a picture or array and find the answer. Complete the sentences.
- Fill in the number model.

1. Five children are playing a game with a deck of 30 cards. How many cards can the dealer give each player?

children	cards per child	cards in all

_____ cards to each player. _____ cards are left over.

Number model: _____ ÷ _____ → _____ R_____

2. The pet shop has 12 puppies in pens. There are 4 puppies in each pen. How many pens have puppies in them?

pens	puppies per pen	puppies in all

_____ pens have puppies in them. _____ puppies are left over.

Number model: _____ ÷ _____ → _____ R_____

3. Tennis balls are sold 3 to a can. Luis buys 15 balls. How many cans is that?

cans	balls per can	balls in all

Luis buys _____ cans.

Number model: _____ ÷ _____ → _____ R_____

LESSON 11•5 Division Number Stories *continued*

4. Eight children share 18 toys equally. How many toys does each child get?

______	______ per ______	______ in all

Each child gets ______ toys.

______ toys are left over.

Number model: ______ ÷ ______ → ______ R______

5. Seven friends share 24 marbles equally. How many marbles does each friend get?

______	______ per ______	______ in all

Each friend gets ______ marbles. ______ marbles are left over.

Number model: ______ ÷ ______ → ______ R______

Try This

6. Tina is storing 20 packages of seeds in boxes. Each box holds 6 packages. How many boxes does Tina need to store all the packages? (Be careful. Think!)

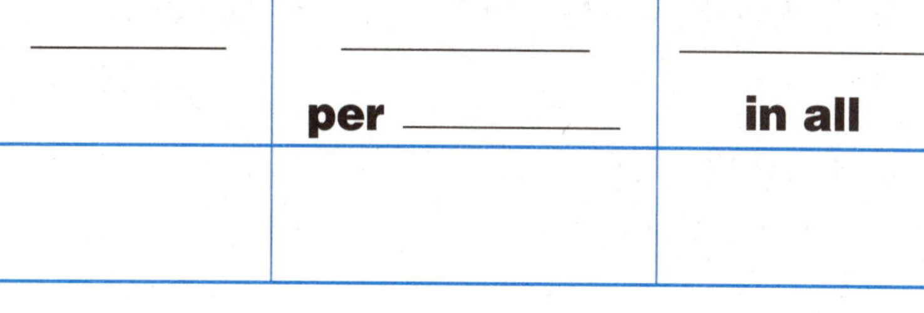

______	______ per ______	______ in all

Tina needs ______ boxes.

Number model: ______ ÷ ______ → ______ R______

Date Time

LESSON 11·5

Math Boxes

1. 8 flower boxes. 4 plants in each box. How many plants?

_____ plants

Fill in the diagram and write a number model.

boxes	plants per box	plants in all

_____ × _____ = _____

MRB 112 113

2.

● ● ● ●
● ● ● ●

_____-by-_____ array

How many dots in all? _____

3. Use the partial-sums algorithm to solve. Show your work. Circle the best answer.

$$\begin{array}{r} 78 \\ +\ 26 \\ \hline \end{array}$$

A. 92 **B.** 52

C. 94 **D.** 104

MRB 30

4. Count by ones.

5,099; _____; _____;

5,102; _____; _____;

5,105; _____; _____

5. Use an inch ruler to find the perimeter of the hexagon.

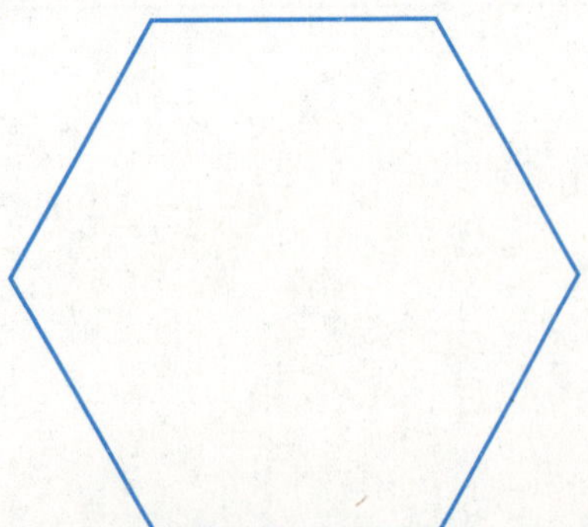

The perimeter is _____ inches.

MRB 68

6. Complete the table.

Rule
×2

in	out
0	
1	
2	
	6
4	
	10

MRB 100–102

Date Time

LESSON 11•6 Multiplication Facts List

I am listing the times _____ facts.

If you are not sure of a fact, draw an array with Os or Xs.

2 × _____ = _____

3 × _____ = _____

4 × _____ = _____

5 × _____ = _____

6 × _____ = _____

7 × _____ = _____

8 × _____ = _____

9 × _____ = _____

10 × _____ = _____

Date Time

LESSON 11·6

Using Arrays to Find Products

Draw an array to help you find each product.
Use Xs to draw your arrays.

1. $2 \times 4 =$ _____

X X X X
X X X X

2. $4 \times 2 =$ _____

3. $6 \times 5 =$ _____

4. $5 \times 6 =$ _____

5. $5 \times 5 =$ _____

6. $2 \times 10 =$ _____

Try This

7. $4 \times 15 =$ _____

Date Time

LESSON 11•6

Math Boxes

1. Subtract. Show your work.

$$\begin{array}{r} 72 \\ -\ 35 \\ \hline \end{array}$$

Unit

2. What is the median (the middle value) for this list of numbers?

______

51, 82, 51, 23, 23, 67

3. This line segment is

______ cm long.

Draw a line segment 4 cm longer.

4. I have a pile of 16 counters.

$\frac{1}{2}$ = ______ counters

$\frac{8}{16}$ = ______ counters

5. Find the rules.

6. Write the mode for this set of numbers (the number that occurs the most often).

______

29, 17, 39, 12, 17

MRB 45

Date Time

LESSON 11·7

Products Table

0×0 = **0**	0×1 =	0×2 =	0×3 =	0×4 =	0×5 =	0×6 =	0×7 =	0×8 =	0×9 =	0×10 =
1×0 =	1×1 = **1**	1×2 =	1×3 =	1×4 =	1×5 =	1×6 =	1×7 =	1×8 =	1×9 =	1×10 =
2×0 =	2×1 =	2×2 = **4**	2×3 =	2×4 =	2×5 =	2×6 =	2×7 =	2×8 =	2×9 =	2×10 =
3×0 =	3×1 =	3×2 =	3×3 = **9**	3×4 =	3×5 =	3×6 =	3×7 =	3×8 =	3×9 =	3×10 =
4×0 =	4×1 =	4×2 =	4×3 =	4×4 = **16**	4×5 =	4×6 =	4×7 =	4×8 =	4×9 =	4×10 =
5×0 =	5×1 =	5×2 =	5×3 =	5×4 =	5×5 = **25**	5×6 =	5×7 =	5×8 =	5×9 =	5×10 =
6×0 =	6×1 =	6×2 =	6×3 =	6×4 =	6×5 =	6×6 = **36**	6×7 =	6×8 =	6×9 =	6×10 =
7×0 =	7×1 =	7×2 =	7×3 =	7×4 =	7×5 =	7×6 =	7×7 = **49**	7×8 =	7×9 =	7×10 =
8×0 =	8×1 =	8×2 =	8×3 =	8×4 =	8×5 =	8×6 =	8×7 =	8×8 = **64**	8×9 =	8×10 =
9×0 =	9×1 =	9×2 =	9×3 =	9×4 =	9×5 =	9×6 =	9×7 =	9×8 =	9×9 = **81**	9×10 =
10×0 =	10×1 =	10×2 =	10×3 =	10×4 =	10×5 =	10×6 =	10×7 =	10×8 =	10×9 =	10×10 = **100**

Date Time

LESSON 11•7

Math Boxes

1. 5 nests with 3 eggs in each. How many eggs in all?

_____ eggs

nests	eggs per nest	eggs in all

_____ × _____ = _____

2. Maria has 9 pairs of shoes in her closet. How many shoes does she have in all?

Draw an array.

_____ × _____ = _____

_____ shoes

3. Solve.

Unit
baby alligators

_____ = 24 + 41

33 + 12 = _____

_____ = 52 + 15

16 + 51 = _____

4. Count by thousands. Fill in the circle next to the best answer.

2,324; _______; 4,324

Ⓐ 3,224 Ⓑ 2,424

Ⓒ 3,324 Ⓓ 3,434

5. Draw a rectangle.
Make two sides
3 inches long.
Make the other two sides
2 inches long.

6. Complete the table.

Rule
×2

in	out
20	
	60
40	

LESSON 11•8 Multiplication/Division Fact Families

Write the fact family for each Fact Triangle.

1.

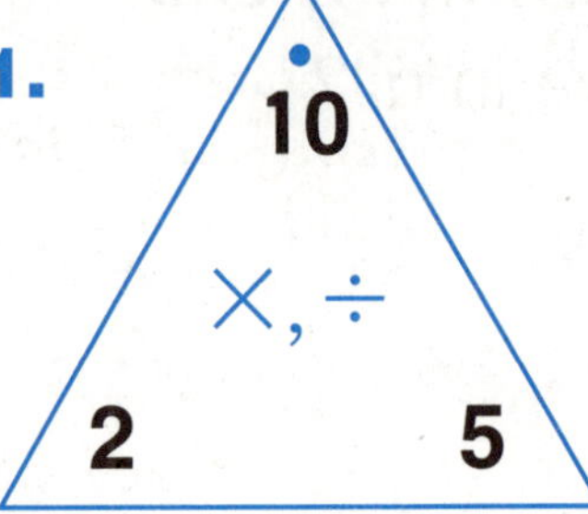

5 × 2 = 10

___ × ___ = ___

10 ÷ 2 = 5

___ ÷ ___ = ___

2.

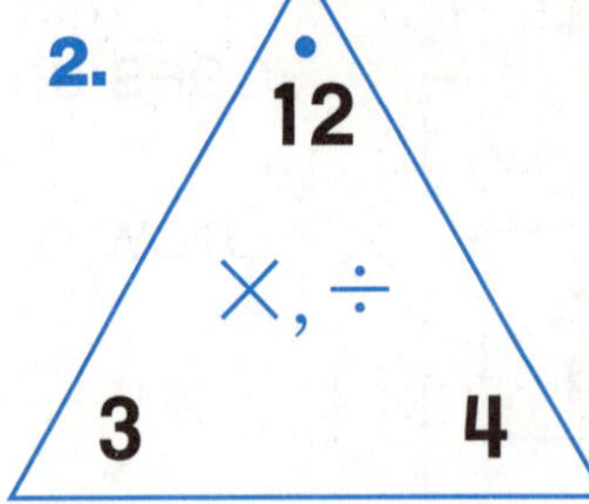

___ × ___ = ___

___ × ___ = ___

___ ÷ ___ = ___

___ ÷ ___ = ___

3.

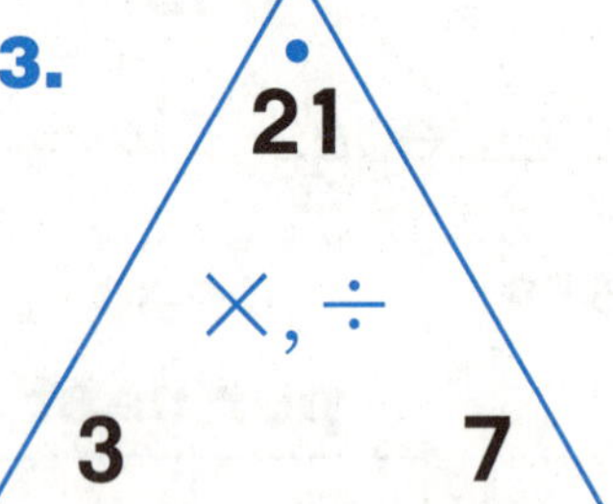

___ × ___ = ___

___ × ___ = ___

___ ÷ ___ = ___

___ ÷ ___ = ___

4.

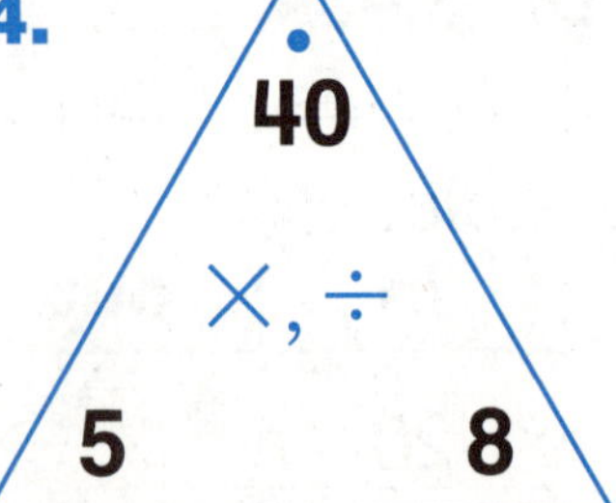

___ × ___ = ___

___ × ___ = ___

___ ÷ ___ = ___

___ ÷ ___ = ___

5.

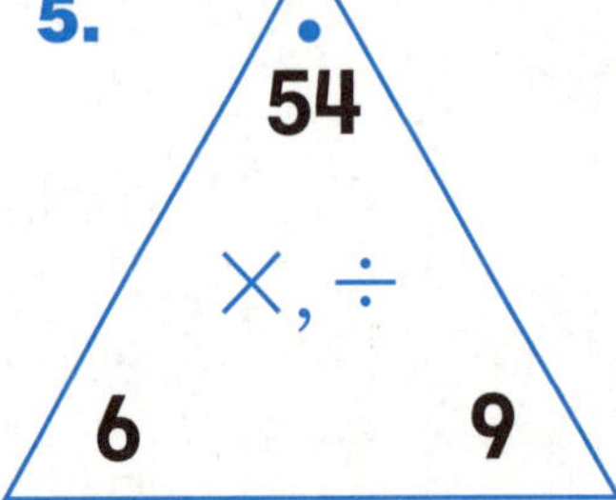

___ × ___ = ___

___ × ___ = ___

___ ÷ ___ = ___

___ ÷ ___ = ___

6.

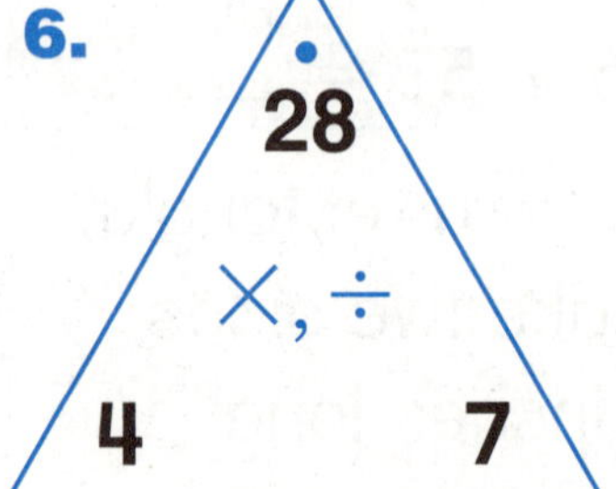

___ × ___ = ___

___ × ___ = ___

___ ÷ ___ = ___

___ ÷ ___ = ___

Date Time

LESSON 11·8

Multiplication and Division with 2, 5, and 10

1. double 8 = _____ _____ = double 9

2 × 4 = _____ _____ = 2 × 1

7 × 2 = _____ _____ = 0 × 2

2. 40 cents = _____ nickels 5 ÷ 1 = _____

15 ÷ 5 = _____ _____ nickels = 25 cents

3. 40 cents = _____ dimes _____ ÷ 10 = 6

20 ÷ 10 = _____ _____ dimes = 90 cents

4. For each multiplication fact, give two division facts in the same fact family.

2 × 6 = 12	12 ÷ 2 = 6	12 ÷ 6 = 2
5 × 9 = 45	_____ ÷ _____ = _____	_____ ÷ _____ = _____
10 × 4 = 40	_____ ÷ _____ = _____	_____ ÷ _____ = _____
3 × 2 = 6	_____ ÷ _____ = _____	_____ ÷ _____ = _____
8 × 5 = 40	_____ ÷ _____ = _____	_____ ÷ _____ = _____
5 × 10 = 50	_____ ÷ _____ = _____	_____ ÷ _____ = _____

LESSON 11·8

Math Boxes

1. Subtract. Show your work.

$$\begin{array}{r} 90 \\ -\ 64 \\ \hline \end{array} \qquad \begin{array}{r} 37 \\ -\ 18 \\ \hline \end{array}$$

Unit

2. What is the median (the middle value) for this list of numbers?

____

50, 31, 41, 42, 41

3. Draw a line segment 3 cm long.

Draw a second line segment 4 cm longer than the first.

Draw a third line segment twice as long as the first.

4. Get 21 counters.

$\frac{1}{3}$ = ________ counters

$\frac{2}{7}$ = ________ counters

$\frac{3}{3}$ = ________ counters

5. Complete the frames.

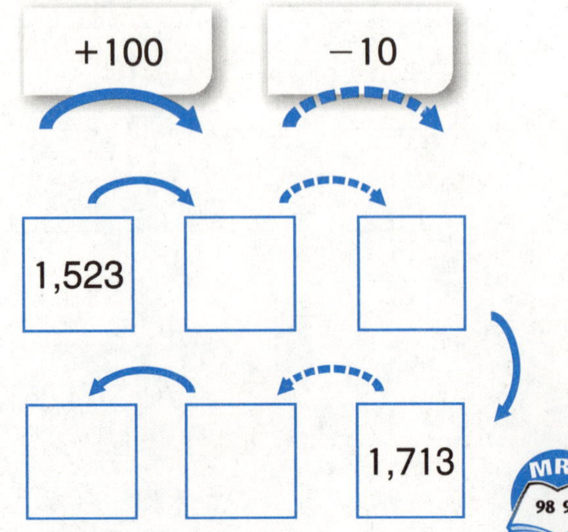

MRB 98 99

6. Find the mode (the number that occurs most often). Choose the best answer.

496, 738, 713, 100, 713

- ⬭ 713
- ⬭ 496
- ⬭ 100
- ⬭ 738

MRB 45

Date Time

LESSON 11•9

Math Boxes

1. Each ladybug has 5 spots. 20 spots in all.

How many ladybugs? ______

Fill in the diagram and write a number model.

ladybugs	spots per ladybug	spots in all

______ × ______ = ______

MRB 112 113

2. Multiply. If you need help, make arrays.

Unit

4 × 5 = ______

6 × 2 = ______

1 × 10 = ______

3. Complete.

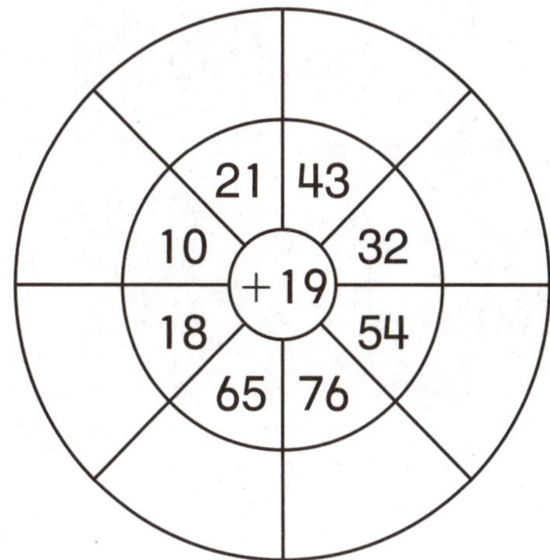

4. Write the number that is 1,000 more than:

7,542 ________

1,837 ________

5,641 ________

8,863 ________

5. Draw a line 1 inch long.

Draw another line twice as long.

6. Find the rule. Complete the table.

in	out
10¢	20¢
15¢	30¢
	50¢
60¢	

Rule

Beat the Calculator

Materials ☐ calculator

Players 3 (Caller, Brain, and Calculator)

Directions

1. The Caller reads fact problems from the Brain's journal—in the order listed on the next page.
2. The Brain solves each problem and says the answer.
3. While the Brain is working on the answer, the Calculator solves each problem using a calculator and says the answer.
4. If the Brain beats the Calculator, the Caller makes a check mark next to the fact in the Brain's journal.

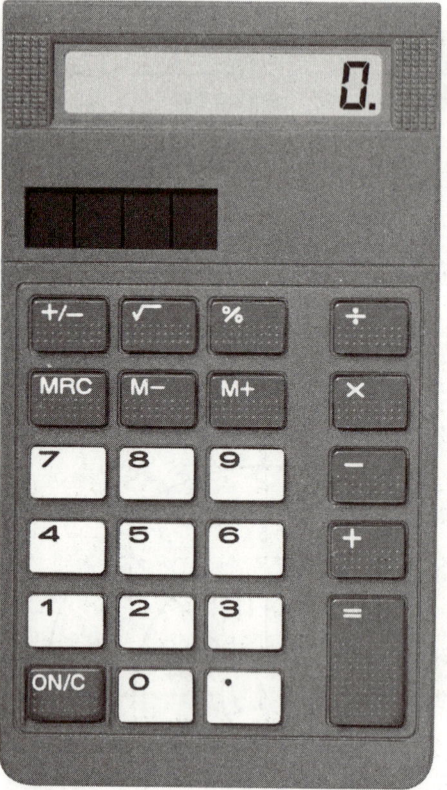

Date Time

Beat the Calculator *continued*

✓	✓	✓	Fact Problem
			2 × 4 = ____
			3 × 5 = ____
			2 × 2 = ____
			4 × 3 = ____
			5 × 5 = ____
			6 × 2 = ____
			6 × 5 = ____
			3 × 3 = ____
			4 × 5 = ____
			3 × 6 = ____

✓	✓	✓	Fact Problem
			7 × 3 = ____
			5 × 2 = ____
			6 × 4 = ____
			2 × 7 = ____
			3 × 2 = ____
			4 × 4 = ____
			4 × 1 = ____
			4 × 7 = ____
			7 × 5 = ____
			0 × 2 = ____

Math Boxes

1. What is the range of this set of numbers (the largest number minus the smallest number)?

______

81, 910, 109, 175

2. Find the mode (the number that occurs most often).

______

183, 56, 618, 56, 215, 56, 183, 56

3. Complete the Fact Triangle. Write the fact family.

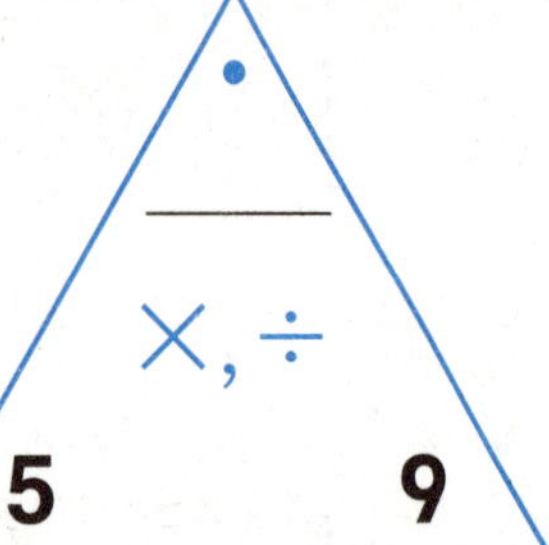

______ × ______ = ______

______ × ______ = ______

______ ÷ ______ = ______

______ ÷ ______ = ______

4. Find the median (the middle number).

______

640, 710, 615, 915, 320

5. Complete the table.

Rule
×2

in	out
100	
	400
1,000	
	4,000

6. Complete the Fact Triangle. Write the fact family.

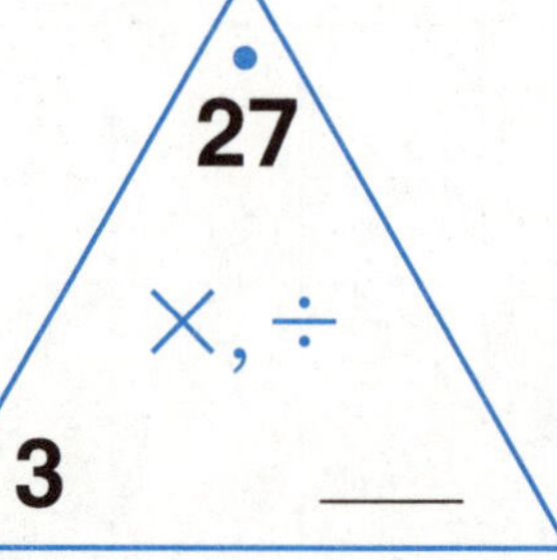

______ × ______ = ______

______ × ______ = ______

______ ÷ ______ = ______

______ ÷ ______ = ______

LESSON 12•1 Review: Telling Time

1. How many hours do clock faces show? _____ hours
2. How long does it take the hour hand to move from one number to the next? ______________________________
3. How long does it take the minute hand to move from one number to the next? ______________________________
4. How many times does the hour hand move around the clock face in one day? _____ times
5. How many times does the minute hand move around the clock face in one day? _____ times

Write the time shown by each clock.

6. 

_____ : _____

7. 

_____ : _____

8. 

_____ : _____

Draw the hour and minute hands to match the time.

9. 

8:00

10. 

6:45

11. 

4:10

LESSON 12•1

Math Boxes

1. Jordan spent \$6.38 on a book and \$1.23 on a magazine. How much did he spend all together? First estimate the costs and the total.

_____ + _____ = _____

Then use partial sums and solve.

2. 1 hour = ____ minutes

$\frac{1}{2}$ hour = ____ minutes

$\frac{1}{4}$ hour = ____ minutes

$\frac{3}{4}$ hour = ____ minutes

$1\frac{1}{2}$ hours = ____ minutes

3. Write a fraction for each shaded part. Put $<$, $>$, or $=$ in the box.

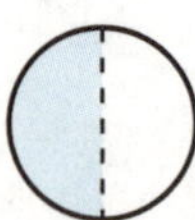

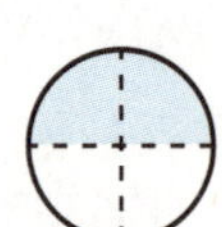

4. I spent \$4.22 at the store and gave the cashier a \$10 bill. How much change should I get?

\$_____

5.

5,401	1,290	632
3,679	890	798

The minimum number is _____.

The maximum number is _____.

6. A pentagon has ___ sides.

A hexagon has ___ sides.

An octagon has ___ sides.

MRB 54

Date Time

LESSON 12•2

Time Before and After

1. It is:

Show the time 20 minutes later.

What time is it?

____ : ____

2. It is:

Show the time 35 minutes later.

What time is it?

____ : ____

3. It is:

Show the time 15 minutes earlier.

What time is it?

____ : ____

4. You pick a time. Draw the hands on the clock.

It is:

Show the time 50 minutes later.

What time is it?

____ : ____

Date Time

LESSON 12·2

Many Names for Times

What time does each clock show? Fill in the ovals next to the correct names.

Example: ● quarter-past 1 ● 15 minutes after 1

O two ten O 5 minutes after 3

● one fifteen

1. O seven fifteen O quarter-to 7

O quarter-to 8 O quarter-past 8

O quarter-past 7

2. O half-past 10 O eleven thirty

O half-past 11 O 30 minutes after 10

O ten thirty

3. O quarter-past 5 O quarter-to 6

O quarter-to 5 O six fifteen

O five forty-five

4. O nine forty O 20 to 8

O 20 to 9 O eight forty

O 40 minutes to 9

Addition and Subtraction Strategies

Add or subtract. Use your favorite addition or subtraction strategy.

1. 53 + 45 **Answer**	**2.** $\begin{array}{r} 36 \\ +\ 48 \\ \hline \end{array}$ **Answer**	**3.** 456 + 17 **Answer**
4. 68 − 24 **Answer**	**5.** $\begin{array}{r} 65 \\ -\ 27 \\ \hline \end{array}$ **Answer**	**6.** 516 − 38 **Answer**

LESSON 12•2

Math Boxes

1. **Favorite Animals of Ms. Lee's Classroom**

Whales	XXX
Manatees	XXXXXX
Walruses	XXXXX
Cubs	XXXXXXXXXX

Which animal was the most popular? ______

2. Solve.

Unit

$(14 - 7) + 4 =$ ____

$14 - (7 + 4) =$ ____

$(12 - 7) + 5 =$ ____

$12 - (7 + 5) =$ ____

3. Write < or >.

2,469 ______ 12,469

60,278 ______ 50,278

25,100 ______ 25,110

MRB 9

4. What item is the shape of a rectangular prism? Circle the best answer.

A doughnut

B can of soup

C book

D ice-cream cone

5.

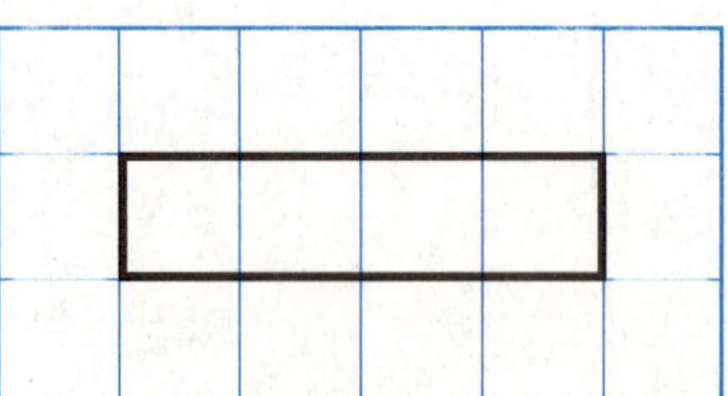

The area is ______ sq cm.

The perimeter is ______ cm.

6. The pet store sold 12 fish. $\frac{1}{2}$ were guppies and $\frac{1}{4}$ were neons. The rest were angelfish. How many of each?

There were ______ guppies.

There were ______ neons.

There were ______ angelfish.

Date Time

Important Events in Communication

For each event below, make a dot on the timeline and write the letter for the event above the dot.

A telephone (1876)

B radio (1906)

C television (1926)

D telegraph (1837)

E CD player (1982)

F copier (1937)

G audiocassette (1963)

H phonograph (1877)

I personal computer (1974)

J movie machine (1894)

K 3-D movies (1922)

L videocassette (1969)

M typewriter (1867)

N FM radio (1933)

A

1830 1840 1850 1860 1870 1880 1890 1900 1910 1920 1930 1940 1950 1960 1970 1980 1990 2000 2010

Date Time

LESSON 12•3

Math Boxes

1. A baseball costs \$3.69. A yo-yo costs \$1.49. You buy both.

Estimate the cost:

______ + ______ = ______

Actual cost:

\$______

2. Naquon grew 2 inches in _____. Circle the best answer.

A 1 day

B 5 minutes

C 1 year

D 24 hours

3. Write four names for $\frac{1}{2}$. Use your Fraction Cards to help.

_____, _____, _____, _____

Write 2 fractions that are:

greater than $\frac{1}{2}$. _____, _____

less than $\frac{1}{2}$. _____, _____

4. I bought a beach ball for \$1.49 and a sand toy for \$3.96. How much change will I get from a \$10 bill?

\$______

5. \$24.06 \$9.99 \$14.98 \$19.99 \$29.83

The maximum is

\$______.

The minimum is

\$______.

6. Draw two polygons with 4 sides.

MRB 52

Date Time

LESSON 12•4

Subtraction Practice

Fill in the unit box. Then, for each problem:

Unit

- Make a ballpark estimate before you subtract.
- Write a number model for your estimate. Then solve the problem. For Problems 1 and 2, use the trade-first algorithm. For Problems 3–6, use any strategy you choose.
- Compare your estimate to your answer.

1. Ballpark estimate: ______ 25 − 18 = ______	**2.** Ballpark estimate: ______ 31 − 22 = ______	**3.** Ballpark estimate: ______ 53 − 29 = ______
4. Ballpark estimate: ______ 87 − 39 = ______	**5.** Ballpark estimate: ______ 148 − 29 = ______	**6.** Ballpark estimate: ______ 177 − 48 = ______

Date Time

LESSON 12•4

Math Boxes

1.

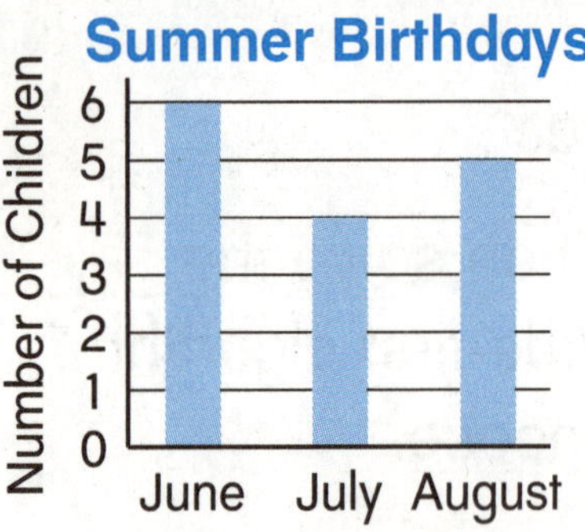

How many birthdays are in June and July? _____

2. Add parentheses to make the number models true.

Unit
children

$18 - 13 - 4 = 9$

$18 - 13 - 4 = 1$

$27 - 6 + 10 = 31$

$4 \times 2 + 3 = 20$

3. Write <, >, or =.

10,000 _____ 9,999

33,231 _____ 30,231

75,679 _____ 75,855

4. Name 3 objects that are shaped like a cone.

____

____

____

MRB 57

5. Solve.

Area: _____ sq cm

Perimeter: _____ cm

6. Mrs. Bell had 30 pennies. She gave $\frac{1}{3}$ of the pennies to Max and $\frac{1}{2}$ of the pennies to Julie.

Max received _____ pennies.

Julie received _____ pennies.

How many pennies did Mrs. Bell have left?

_____ pennies

Date Time

LESSON 12•5

Related Multiplication and Division Facts

Solve each multiplication fact. Use the fact triangles to help you.

Then use the three numbers to write two division facts.

1. $3 \times 7 =$ 21

21 ÷ 7 = 3

21 ÷ 3 = 7

2. $3 \times 8 =$ ____

____ ÷ ____ = ____

____ ÷ ____ = ____

3. $3 \times 9 =$ ____

____ ÷ ____ = ____

____ ÷ ____ = ____

4. $4 \times 7 =$ ____

____ ÷ ____ = ____

____ ÷ ____ = ____

5. $4 \times 8 =$ ____

____ ÷ ____ = ____

____ ÷ ____ = ____

6. $4 \times 9 =$ ____

____ ÷ ____ = ____

____ ÷ ____ = ____

7. $5 \times 7 =$ ____

____ ÷ ____ = ____

____ ÷ ____ = ____

8. $5 \times 8 =$ ____

____ ÷ ____ = ____

____ ÷ ____ = ____

LESSON 12•5 Addition Card Draw Directions

Materials
- ☐ score sheet from *Math Masters,* p. 446
- ☐ 4 each of number cards 1–10
- ☐ 1 each of the number cards 11–20
- ☐ slate or scratch paper

Players 2

Skill Add 3 numbers

Object of the Game To get the higher total

Directions

Shuffle the cards and place the deck with the numbers facing down. Take turns.

1. Draw the top 3 cards from the deck.
2. Record the numbers on the score sheet. Put the 3 cards in a separate pile.
3. Find the sum. Use your slate or paper to do the computation.

After 3 turns:

4. Check your partner's work. Use a calculator.
5. Find the total of the 3 answers. Write the total on the score sheet. The player with the higher total wins.

LESSON 12•5

Math Boxes

1. Write the number that is

	10 more	100 less
368	______	______
4,789	______	______
40,870	______	______
1,999	______	______

2. How many days per week?

______

How many minutes per hour?

______

How many hours per day?

______

How many weeks per year?

______

MRB 86

3. Shade $\frac{1}{2}$ of the shape. Write the equivalent fraction.

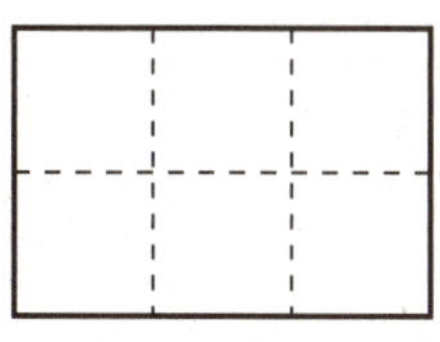

______

4. Fill in the table.

in	out
1	
6	
10	
	80
	250

Rule

$\times 10$

5. The time is

____ : ____.

20 minutes later will be

____ : ____.

15 minutes earlier was

____ : ____.

6. Solve.

Unit

____ − 23 = 17

60 − ____ = 28

49 = ____ − 21

54 = 80 − ____

Date Time

LESSON 12·6

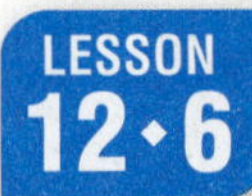

Animal Bar Graph

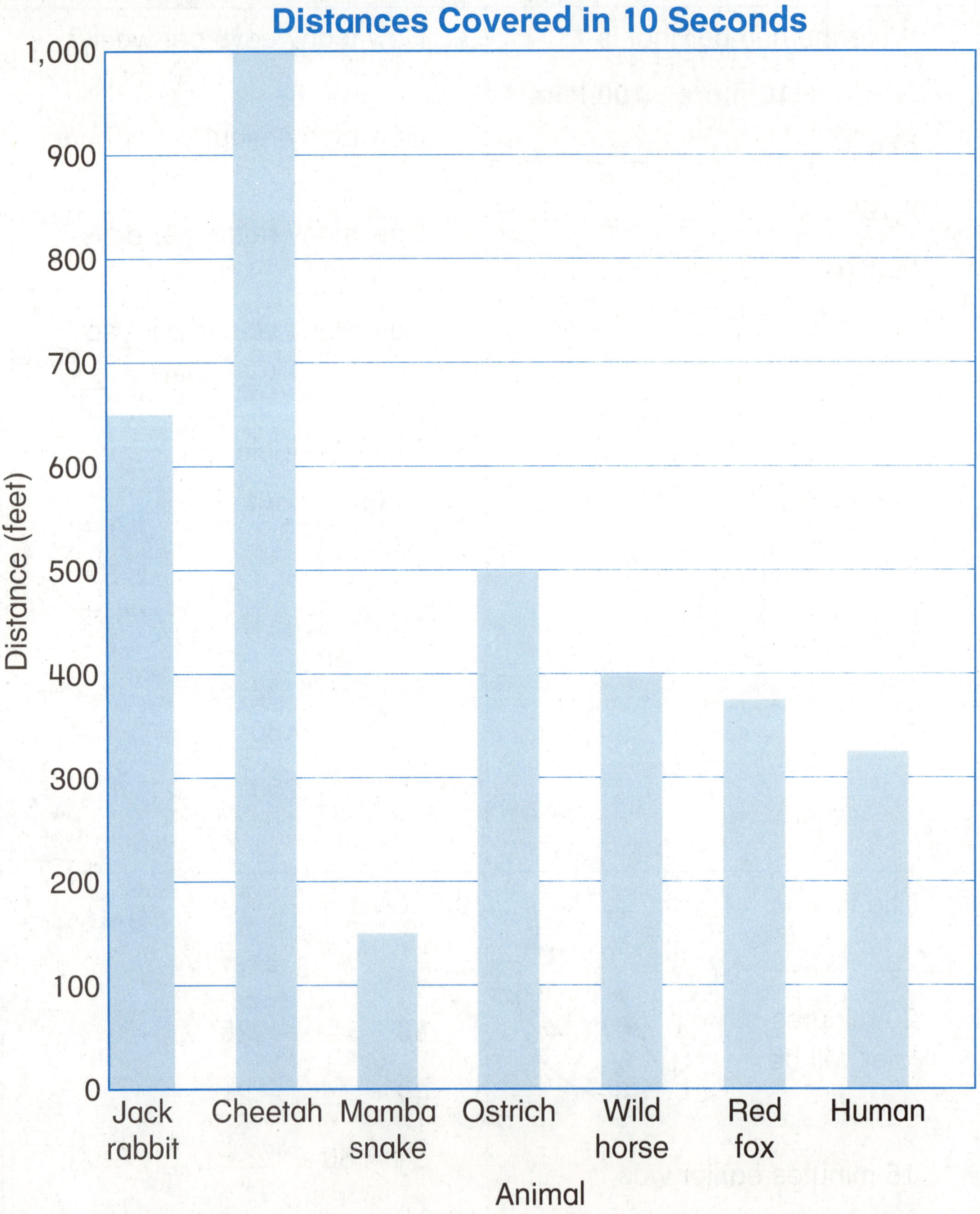
Distances Covered in 10 Seconds
1,000
900
800
700
600
500
400
300
200
100
0
Distance (feet)
Jack rabbit
Cheetah
Mamba snake
Ostrich
Wild horse
Red fox
Human
Animal

Interpreting an Animal Bar Graph

1. In the table, list the animals in order of distance covered in 10 seconds. List the animals from the greatest distance to the least distance.

Animal	Distance
greatest: ______	______ ft
______	______ ft
______	______ ft
______	______ ft
______	______ ft
______	______ ft
least: ______	______ ft

Distances Covered in 10 Seconds

2. Find the middle value of the distances. The middle value is also called the **median.**

   The median is ______ feet.

3. The longest distance is

   ______ feet.

   The shortest distance is

   ______ feet.

4. Fill in the comparison diagram with the longest distance and the shortest distance.

   Quantity

   Quantity

   ______ Difference

5. Find the difference between the longest and shortest distances. The difference between the largest and smallest numbers in a data set is called the **range.**

   The range is ______ feet.

Date Time

LESSON 12•6

Math Boxes

1. Complete the bar graph.

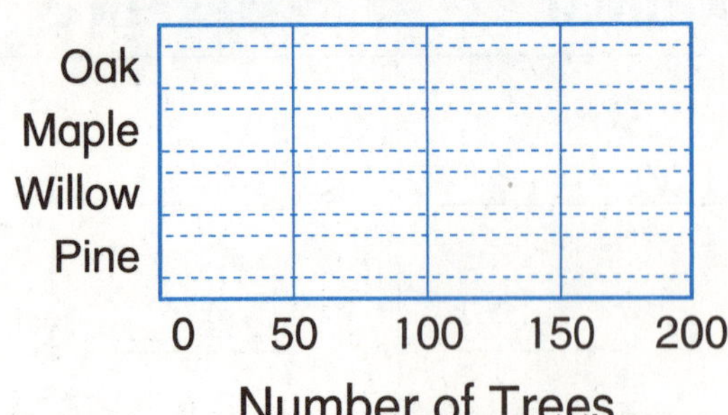

Oak: 100 Maple: 200

Willow: 50 Pine: 150

2. Put parentheses to make each number model true.

$21 = 39 - 10 - 8$

$4 \times 3 + 7 = 40$

$3 \times 5 + 2 = 17$

3. Write <, >, or =.

20,739 _____ 24,596

10,670 _____ 6,670

15,139 _____ 15,264

MRB 9

4. Name the shape.

basketball __________

shoe box

____

paper towel roll __________

5. Draw a shape with an area of 12 square centimeters.

MRB 69

6. A shark swam 80 miles. A seal swam $\frac{1}{2}$ as far as the shark.

How far did the seal swim?

_____ miles

A dolphin swam twice as far as the shark. How far did the dolphin swim?

_____ miles

Date Time

LESSON 12•7

Height Changes

The data in the table show the height of 30 children at ages 7 and 8. Your teacher will show you how to make a line plot for the data.

Student	Height	
	7 Years	8 Years
#1	120 cm	123 cm
#2	132 cm	141 cm
#3	112 cm	115 cm
#4	122 cm	126 cm
#5	118 cm	122 cm
#6	136 cm	144 cm
#7	123 cm	127 cm
#8	127 cm	133 cm
#9	115 cm	120 cm
#10	119 cm	125 cm
#11	122 cm	126 cm
#12	103 cm	107 cm
#13	129 cm	136 cm
#14	124 cm	129 cm
#15	109 cm	110 cm

Student	Height	
	7 Years	8 Years
#16	118 cm	122 cm
#17	120 cm	126 cm
#18	141 cm	148 cm
#19	122 cm	127 cm
#20	120 cm	126 cm
#21	120 cm	124 cm
#22	136 cm	142 cm
#23	115 cm	118 cm
#24	122 cm	130 cm
#25	124 cm	129 cm
#26	123 cm	127 cm
#27	131 cm	138 cm
#28	126 cm	132 cm
#29	121 cm	123 cm
#30	118 cm	123 cm

Height Changes *continued*

Use the line plot your class made to make a frequency table for the data.

Frequency Table	
Change in Height	**Number of Children**
0 cm	
1 cm	
2 cm	
3 cm	
4 cm	
5 cm	
6 cm	
7 cm	
8 cm	
9 cm	
10 cm	

LESSON 12•7

Height Changes *continued*

1. Make a bar graph of the data in the frequency table.

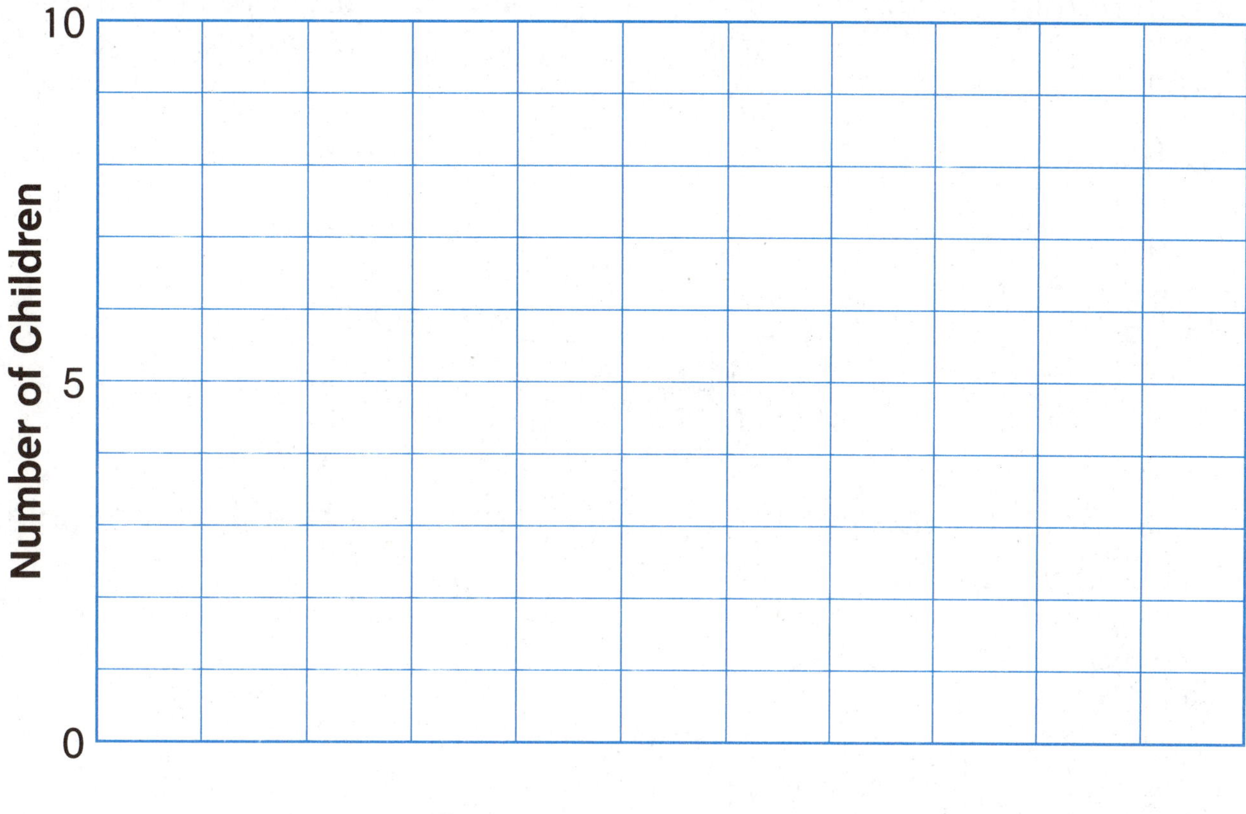

2. The minimum is _____ centimeter(s).

3. The maximum is _____ centimeter(s).

4. The median (the middle value) for the height change data is _____ centimeter(s).

5. The mode (the height change that occurred most often) is _____ centimeter(s).

6. The range is _____ centimeter(s).

Date Time

LESSON 12•7

Math Boxes

1. Write another name for each number.

50 tens = ________

32 hundreds = ________

6,240 = 624 ________

12,000 = 12 ________

2. Jim ate dinner in ________. Fill in the circle next to the best answer.

Ⓐ 2 months

Ⓑ 20 minutes

Ⓒ 2 years

Ⓓ 2 weeks

3. Cross out the fractions that do not belong.

$\frac{1}{2}$

$\frac{2}{3}$, $\frac{3}{5}$, $\frac{4}{8}$,
$\frac{6}{12}$, $\frac{6}{8}$, $\frac{5}{10}$, $\frac{1}{4}$

4. Use counters. Fill in the table.

in	out
4	
9	
7	
	20
	40

Rule
× 4

5. Write the time in hours and minutes.

10 minutes past 12 ___:___

quarter to 11 ___:___

half-past 7 ___:___

25 minutes to 8 ___:___

6. Trade first. Then subtract.

$5.44
− $0.29

$5.44
− $3.29

MRB
34 35

LESSON 12·8

Math Boxes

1. Write the number. Use your Place-Value Book if you need help.

3 tens = ______

33 tens = ______

333 tens = ________

2. Match.

1 day	14 days
3 days	48 hours
2 weeks	24 hours
2 days	72 hours

3. Write the fractions.

△△△△
△△△△
△△△△

________ or ________

4. Fill in the table.

Rule
×3

in	out
0	
1	
2	
3	
	12
	30

5. Cross out names that don't belong.

6:15

six fifteen, quarter to 7,

quarter past 6,

15 minutes before 6,

15 minutes after 6

6. Solve.

Unit

$$\begin{array}{r} 687 \\ -\ 409 \\ \hline \end{array} \qquad \begin{array}{r} 569 \\ -\ 372 \\ \hline \end{array}$$

Date Time

Table of Equivalencies

Weight	
kilogram	1,000 g
pound	16 oz
ton	2,000 lb
1 ounce is about 30 g	

Length	
kilometer	1,000 m
meter	100 cm or 10 dm
decimeter	10 cm
centimeter	10 mm
foot	12 in.
yard	3 ft or 36 in.
mile	5,280 ft or 1,760 yd
10 cm is about 4 in.	

Time	
year	365 or 366 days
year	about 52 weeks
year	12 months
month	28, 29, 30, or 31 days
week	7 days
day	24 hours
hour	60 minutes
minute	60 seconds

<	*is less than*
>	*is more than*
=	*is equal to*
=	*is the same as*

Money		
	1¢, or $0.01	(P)
	5¢, or $0.05	(N)
	10¢, or $0.10	(D)
	25¢, or $0.25	(Q)
	100¢, or $1.00	$1

Abbreviations	
kilometers	km
meters	m
centimeters	cm
miles	mi
feet	ft
yards	yd
inches	in.
tons	T
pounds	lb
ounces	oz
kilograms	kg
grams	g
decimeters	dm
millimeters	mm
pints	pt
quarts	qt
gallons	gal
liters	L
milliliters	mL

Capacity
1 pint = 2 cups
1 quart = 2 pints
1 gallon = 4 quarts
1 liter = 1,000 milliliters

Date Time

Notes

Date Time

Notes

Date Time

LESSON **8·5**

Fraction Cards

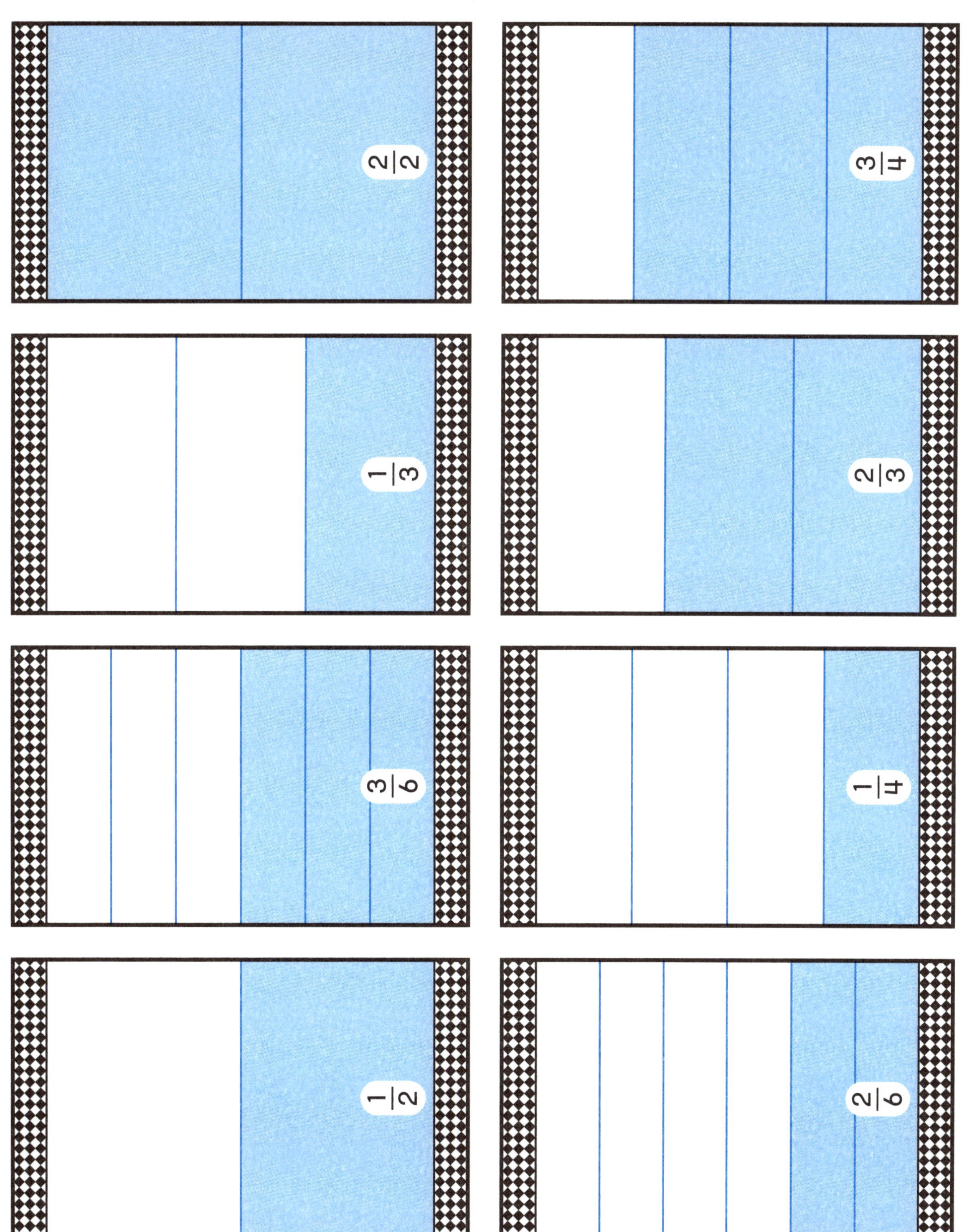

Date Time

Fraction Cards

$\frac{3}{4}$ $\frac{2}{2}$

$\frac{2}{3}$ $\frac{1}{3}$

$\frac{1}{4}$ $\frac{3}{6}$

$\frac{2}{6}$ $\frac{1}{2}$

Date Time

LESSON 8•5

Fraction Cards

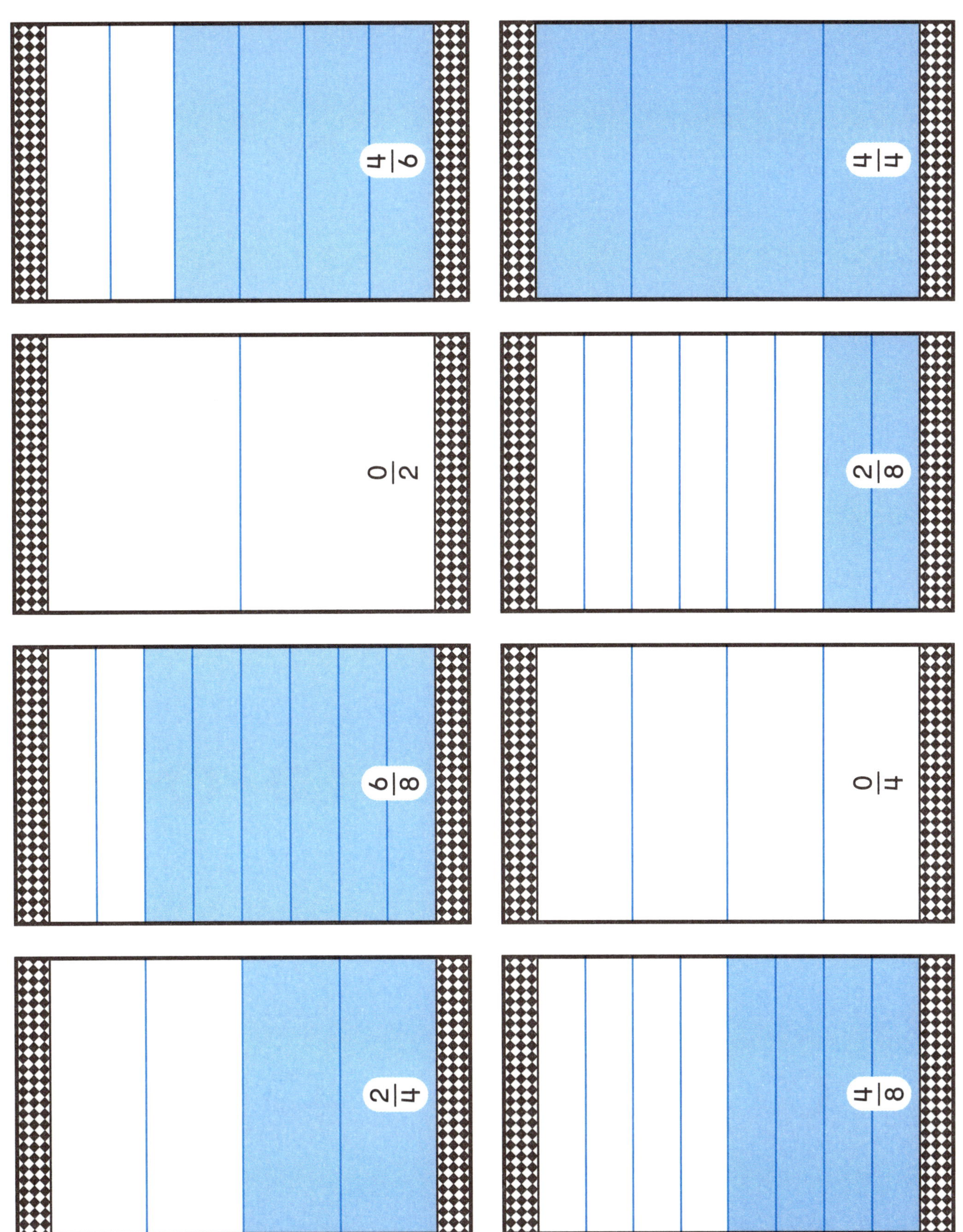

Date Time

Fraction Cards

$\frac{4}{4}$ $\frac{4}{6}$

$\frac{2}{8}$ $\frac{0}{2}$

$\frac{0}{4}$ $\frac{6}{8}$

$\frac{4}{8}$ $\frac{2}{4}$

Date Time

×, ÷ Fact Triangles 1

2 4 8 4

×, ÷ 6 ×, ÷ ×, ÷

2 2 3 2

4 3

2 ×, ÷ 3

×, ÷ ×, ÷

5 10 12 9 3

4 16 20 5

×, ÷ 15 ×, ÷

×, ÷

4 3 5 4

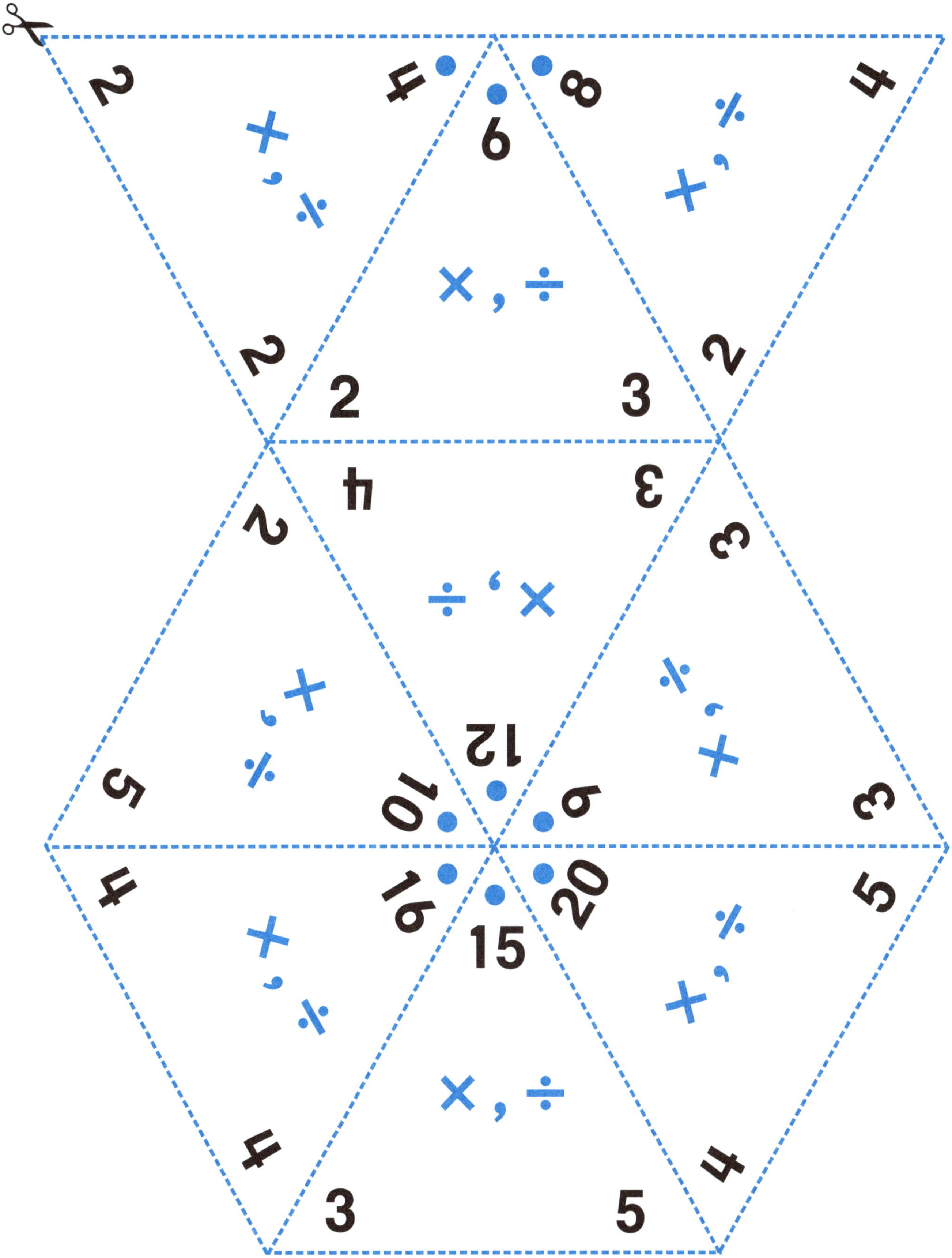

Date Time

×, ÷ Fact Triangles 2

14 ● ×, ÷ 2 7

12 ● ×, ÷ 2 6

35 ● ×, ÷ 7 5

18 ● ×, ÷ 3 6

21 ● ×, ÷ 7 3

24 ● ×, ÷ 6 4

25 ● ×, ÷ 5 5

30 ● ×, ÷ 5 6

28 ● ×, ÷ 7 4

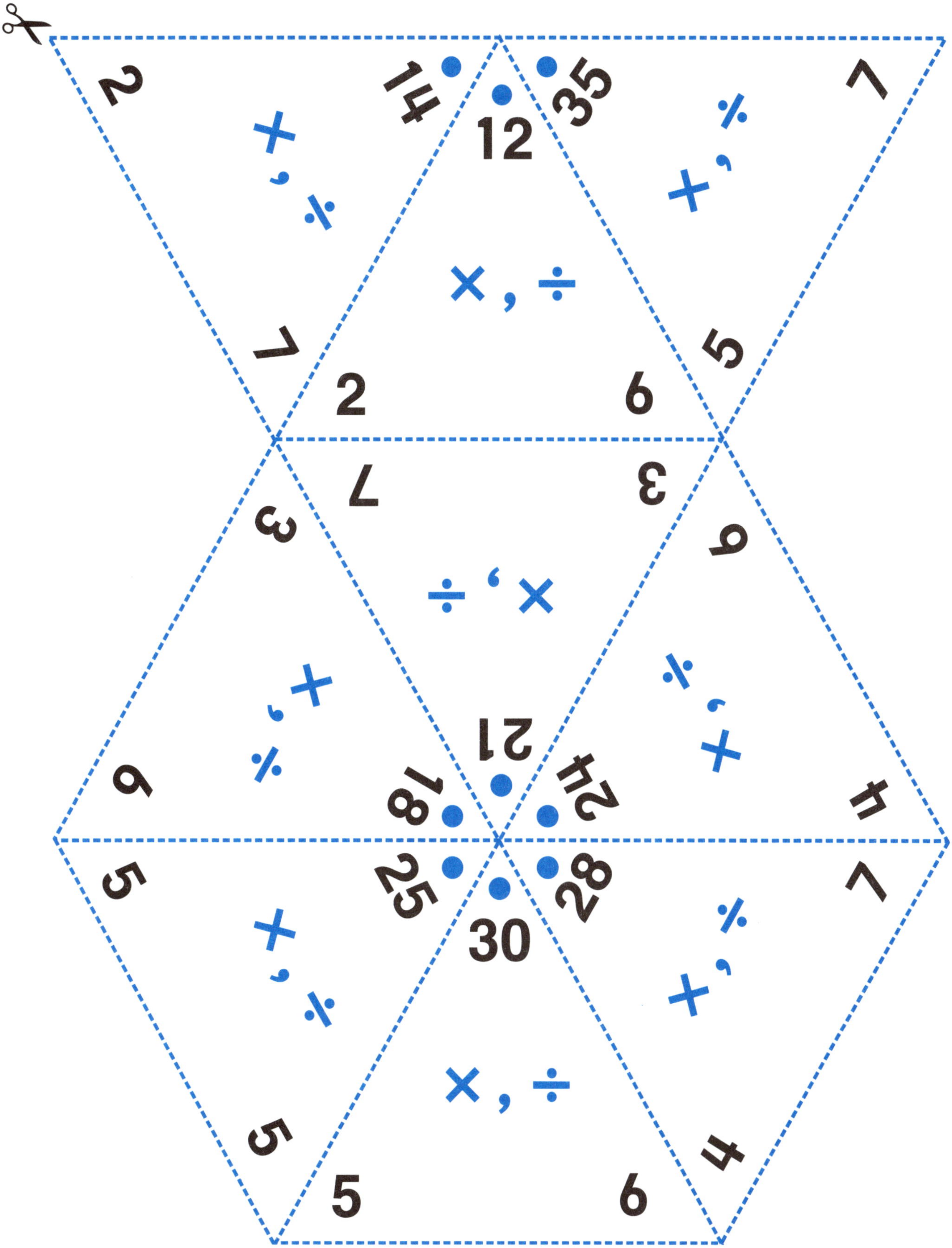

Date Time

×, ÷ Fact Triangles 3

✂

9 36 • 9 ×, ÷

16 • 2 8 ×, ÷

45 • 9 5 ×, ÷

6 4 ×, ÷ 36 •

3 8 24 • ×, ÷

8 4 32 • ×, ÷

3 9 27 • ×, ÷

40 • 5 8 ×, ÷

9 2 18 • ×, ÷

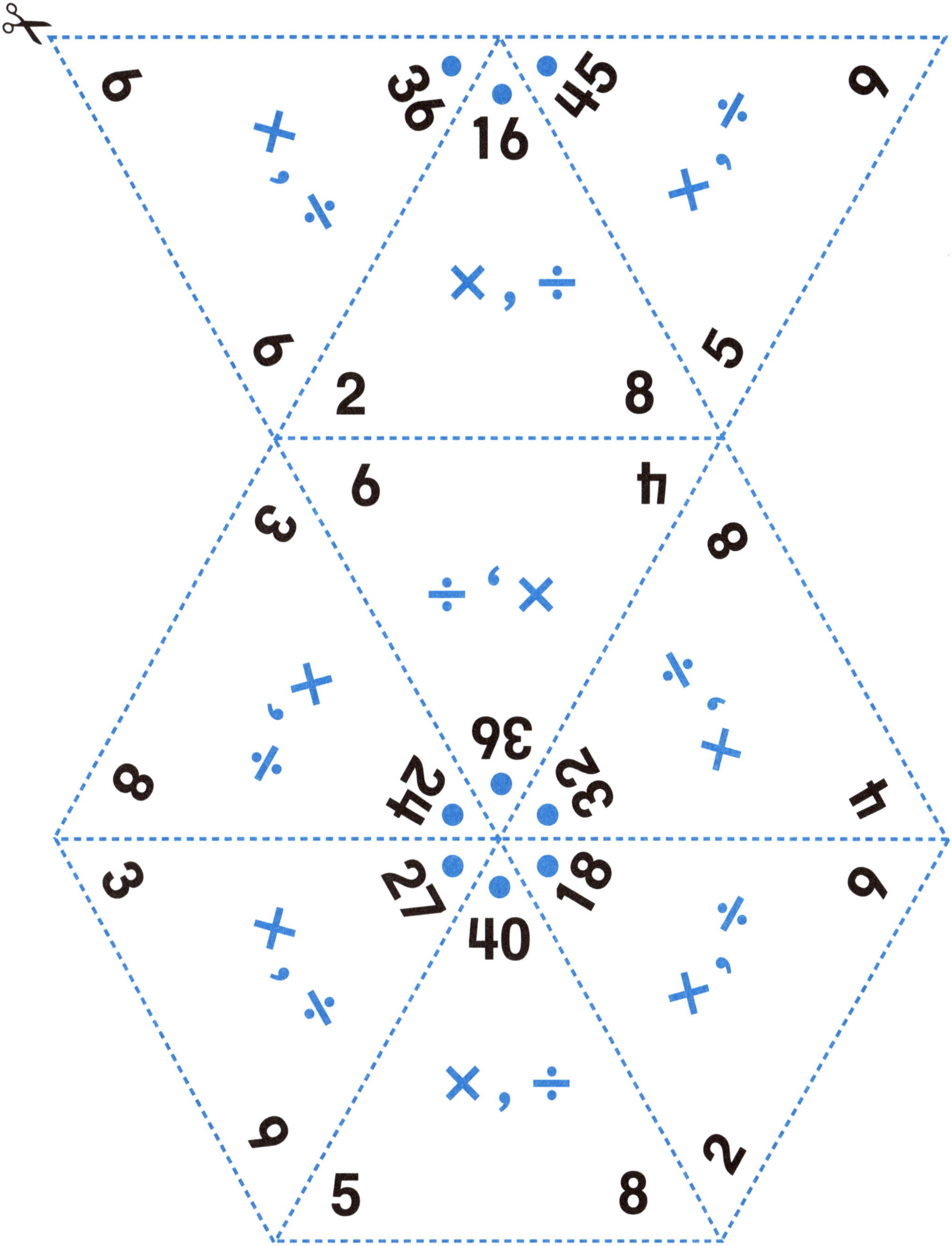

Date Time

LESSON 11·9

×, ÷ Fact Triangles 4

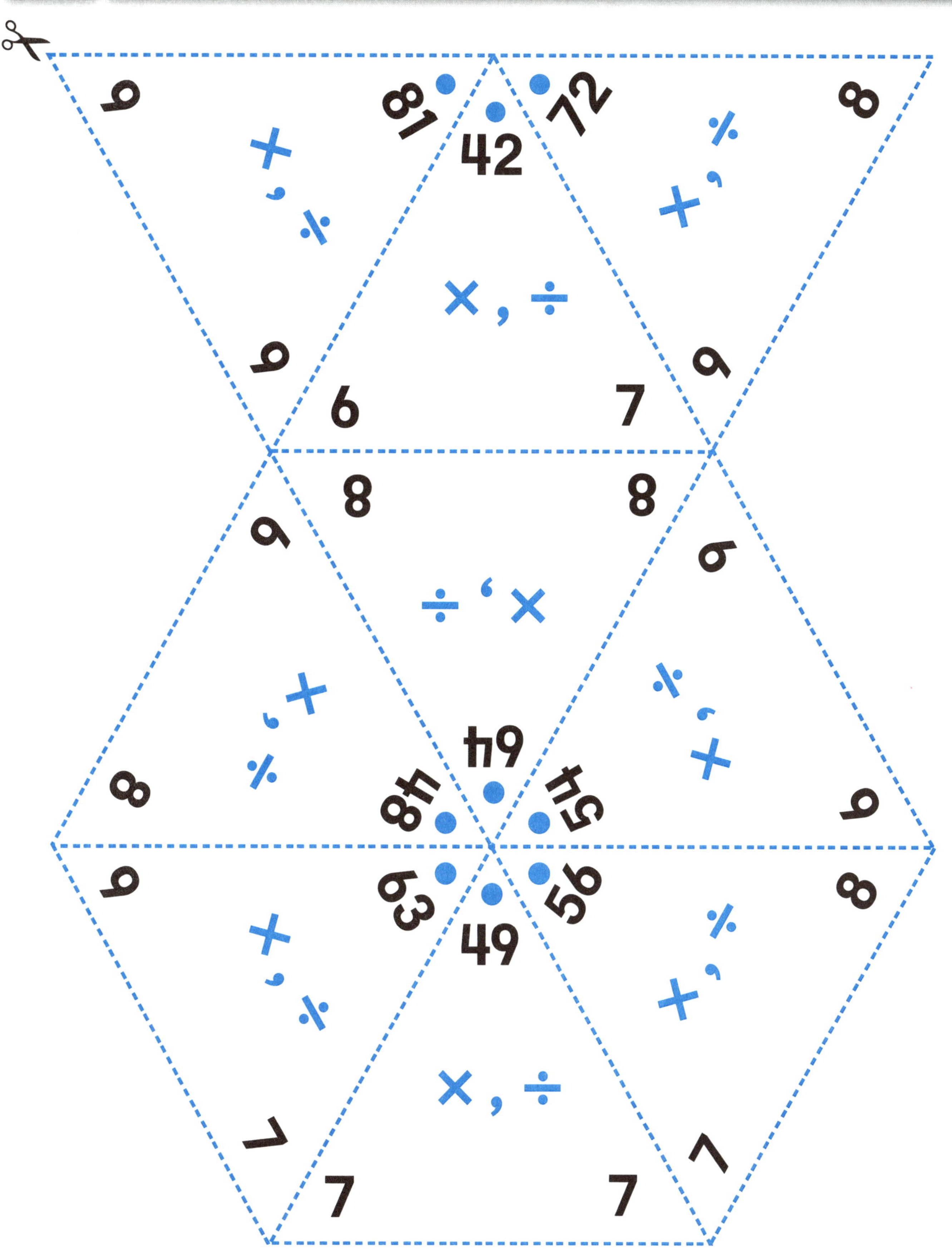